Reservoir Model Design

Reservoir Model Design

Editor
_____

Manoj Negi

Reservoir Model Design

Edited by **Manoj Negi**

Printed in 2017

ISBN: 978-1-68117-154-8

Library of Congress Control Number: 2015936567

© 2016 by
SCITUS Academics LLC,
616, Corporate Way, Suite 2, 4766,
Valley Cottage, NY 10989

www.scitusacademics.com

Notice

Reasonable efforts have been made to publish reliable data and views articulated in the chapters are those of the individual contributors, and not necessarily those of the editors or publishers. Editors or publishers are not responsible for the accuracy of the information in the published chapters or consequences of their use. The publisher believes no responsibility for any damage or grievance to the persons or property arising out of the use of any materials, instructions, methods or thoughts in the book. The editors and the publisher have attempted to trace the copyright holders of all material reproduced in this publication and apologize to copyright holders if permission has not been obtained. If any copyright holder has not been acknowledged, please write to us so we may rectify.

Contents

Preface

This book presents practical advice and ready to use tips on the design and construction of subsurface reservoir models. The design elements cover rock architecture, petrophysical property modelling, multi-scale data integration, upscaling and uncertainty analysis. Philip Ringrose and Mark Bentley share their experience, gained from over a hundred reservoir modelling studies in 25 countries covering clastic, carbonate and fractured reservoir types. The intimate relationship between geology and fluid flow is explored throughout, showing how the impact of fluid type, production mechanism and the subtleties of single- and multi-phase flow combine to influence reservoir model design.

Editor

Optimization of Multiple Hydraulically Fractured Horizontal Wells in Unconventional Gas Reservoirs

Wei Yu and Kamy Sepehrnoori

Department of Petroleum and Geosystems Engineering, the University of Texas at Austin, Austin, TX 78712, USA

ABSTRACT

Accurate placement of multiple horizontal wells drilled from the same well pad plays a critical role in the successful economical production from unconventional gas reservoirs. However, there are high cost and uncertainty due to many inestimable and uncertain parameters such as reservoir permeability, porosity, fracture spacing, fracture half-length, fracture conductivity, gas desorption, and well spacing. In this paper, we employ response surface methodology to optimize multiple horizontal well placement to maximize Net Present Value (NPV) with numerically modeling multistage hydraulic fractures in combination with economic analysis. This paper demonstrates the accuracy of numerical modeling of multistage hydraulic fractures for actual Barnett Shale production data by considering the gas desorption effect. Six uncertain parameters, such as permeability, porosity, fracture spacing, fracture half-length, fracture conductivity, and distance between two neighboring wells with a reasonable range based on Barnett Shale information, are used to fit a response surface of NPV as the objective function and to finally identify the optimum design under conditions of different gas prices based on NPV maximization. This integrated approach can contribute to obtaining the optimal drainage area around the wells by optimizing well placement and hydraulic fracturing treatment design and provide insight into hydraulic fracture interference between single well and neighboring wells.

INTRODUCTION

The combination of horizontal drilling and multistage hydraulic fracturing technology has made possible the current flourishing gas production from shale gas reservoirs in the United States, as well as the global fast growing investment in shale gas exploration and development. Multiple transverse hydraulic fractures are generated when all wellbores are drilled in the direction of the minimum horizontal stress. Maximizing the total stimulated reservoir volume (SRV) plays a major role in successful economic gas production. The unprecedented growth of shale reservoirs has brought a new perspective and focus to the optimization of multiwell placement in the same pad. Drilling multiple horizontal wells from a single pad has increasingly become a common approach for developing shale reservoirs due to significant cost, time, and environmental savings. The surface footprint is reduced greatly by drilling multi-well from the same pad due to minimizing the number of surface locations required while increasing the bottom hole contact of the shale resource [1]. Zipper fracturing (zipper-frac) and simultaneous fracturing (simul-frac) [2], where two adjacent horizontal wells are hydraulically fractured alternatingly and simultaneously, respectively, are two commonly used hydraulic fracturing techniques to stimulate multi-well from the same pad. Although hydraulic fractures improve gas production from shale gas wells, the cost of operation is expensive. Long laterals require greater volume of liquids and proppants which contribute to greater cost [3]. The well economics is also sensitive to well performance and natural gas price due to higher drilling and completion costs. Therefore, optimizing well parameters such as well number and well distance in conjunction with hydraulic fracture parameters, such as fracture spacing and fracture half-length based on economic analysis, are very important, especially in the current environment of low natural gas prices.

Optimization of multi-well placement is primarily valuable for overall project economic viability and minimizing the risks of well collision in shale gas reservoirs. Closer well placement will result in stimulated reservoir volume intersects, leading to well competition and penalizing overall production [5]. However, only recently there have been limited studies of optimizing fracturing design together with multi-well placement simultaneously. Esmaili et al. [6] defined three types of horizontal well based on the number of neighboring wells for sharing drainage area when drilling multiple wells from a pad. Rafiee et al. [7] proposed a new design for two horizontal wells by the modification of the traditional zipper-frac and demonstrated this new design maximized reservoir contact and improved well performance when compared to the original zipper-frac design from both rock mechanics and fluid production aspects. However, such a design assumed a larger fracture spacing of 500 ft and did not optimize well placement and fracture spacing simultaneously to obtain the optimal design for economic gas production. Díaz de Souza et al. [8] did sensitivity studies of three different wells placement with 2, 3, and 4 horizontal wells within the same stimulation volume in the Haynesville Shale to obtain the optimal well spacing. They stated that four horizontal wells with a well distance of 660 ft was a near-optimal solution for this reservoir. However, critical parameters for developing a play economically, such as fracture spacing and fracture half-length, have not been considered for optimization based on optimal well spacing. Harpel et al. [9] reported that well spacing becomes tighter in parts of the Fayetteville Shale from 600 ft to 400 ft or 300 ft, leading to fluid and proppant volume reductions, while further optimization of stimulation design, especially fracture half-length, is very much required for future development. Ramakrishnan et al. [10] suggested that it is particularly challenging to optimize the stimulation treatment in a multi-well design drilled with few hundreds feet of spacing between one another to maximize coverage around each horizontal well, because the optimization process involves perforation and stage placements, sequential stimulation of these wells, fluid and proppant schedules, treatment rates, and application of diversion technology in an effort to achieve effective stimulation along these wells and between the wells. Hence, a detailed study and a comprehensive approach for optimization of fracturing design and multi-well placement are still significantly necessary.

In this paper, we employed response surface methodology to build the response surface in terms of NPV with six parameters such as reservoir porosity, permeability, fracture half-length, fracture conductivity, fracture spacing, and well distance from Barnett Shale, to obtain the best economic scenario for a given range of these influential parameters. The effect of gas desorption is integrated in the numerical modeling of multistage hydraulic fractures. The impact of different gas prices is also taken into account for the optimization process. The goal of this work is to provide insights into the effective exploitation of shale gas reservoirs via optimization of the fracturing design and multiple wells placement simultaneously.

SHALE GAS RESERVOIR MODELING

Given the complex nature of hydraulic fracture growth and the very low permeability of the matrix rock in shale gas reservoirs, coupled with the predominance of horizontal completions, reservoir simulation is the preferred method to predict and evaluate well performance [11–13]. Local grid refinement with logarithmic cell spacing is used in the simulation to accurately model flow from the shale to the fracture, that is, properly incorporate the transient flow behavior from the matrix to the fracture. In a block, the hydraulic fracture is explicitly modeled; moreover, the matrix is described as some subcells whose size increases logarithmically, while moving away from the hydraulic fracture to properly simulate the large pressure drop between the matrix and the fracture. In addition, a dual permeability grid is used to allow simultaneous matrix-to-matrix and fracture-to-fracture flows. This method can accurately and efficiently model transient gas production from hydraulic fractures of the horizontal wells in shale gas reservoirs

[14, 15]. The reservoir is assumed to be homogeneous and the fractures evenly spaced, with stress-independent porosity and permeability. It is assumed that there is no water flow in the reservoir modeling of shale gas. In our simulation, gas is only flowing into the wellbore through the hydraulic fractures, that is, no matrix-wellbore communication. The turbulent gas flow due to high gas flow rate in hydraulic fractures is modeled as non-Darcy flow. The non-Darcy Beta factor, used in the Forchheimer number, is determined using a correlation proposed by Evans and Civan [16] as follows:

$$\beta_{(f)} = \frac{1.485E9}{K^{1.021}},$$

(1)

where the unit of K is md and the unit of is ft^{-1}. The $\beta_{(f)}$ correlation was obtained using over 180 data points including those for propped fractures and was found to match the data very well with the correlation coefficient of 0.974 [14]. This equation is implemented into the numerical model and used for accounting for non-Darcy flow in hydraulic fractures. Figure 1 is a diagram of typical shale gas completion design with a multistage hydraulic fracture treatment, which illustrates several important geometric fracture parameters, such as outer fractures, inner fractures, fracture spacing, and fracture half-length.

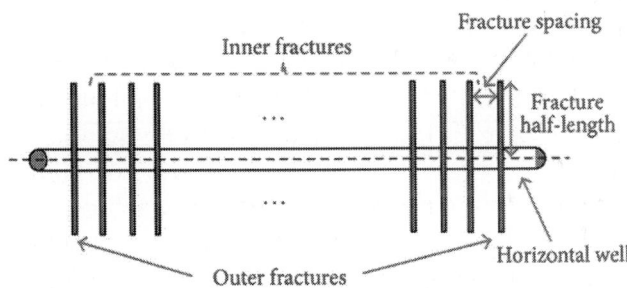

Figure 1: A sketch of multiple hydraulic fractured horizontal shale gas well.

ECONOMIC MODEL

NPV is one of the most common methods used to evaluate the economic viability of investing a project. It is referred to the sum of all cash flows discounted to a specific point in time at the investor's minimum discount rate. The correlation between the present value P and the future value F is

$$P = \frac{F}{(1 + i)^n},$$

(2)

where i is the currency escalation rate or interest rate; n is the number of periods.

The NPV is calculated using the following expression:

$$NPV = \sum_{j=1}^{n} \frac{(V_F)_j}{(1 + i)^j} - \sum_{j=1}^{n} \frac{(V_o)_j}{(1 + i)^j}$$
$$- \left(FC + \sum_{k=1}^{N} (C_{well} + C_{fracture}) \right),$$

(3)

where V_F is future value of production revenue for a fracture reservoir; V_o is future value of production revenue for an unfractured reservoir; FC is the total fixed cost; C_{well} is the cost of one horizontal well; $C_{fracture}$ is the cost of hydraulic fracture in a horizontal well; N is the number of horizontal wells. The costs of well and fracture used in the economic analysis are based on the work of Schweitzer and Bilgesu [17], as shown in Table1.

Table 1: Economic data for NPV calculation

Horizontal well length (ft)	Cost ($)	Fracture half-length per stage (ft)	Cost ($)	Parameter	Value
1,000	2,000,000	250	100,000	Gas Price	$3, 4, 5/ MSCF

2,000	2,100,000	500	125,000	Interest rate	10.0%
3,000	2,200,000	750	150,000	Royalty tax	12.5%
4,000	2,300,000	1,000	175,000		

LANGMUIR ISOTHERM

Gas shales are organic-rich formations. Gas storage in the shale is mainly divided into free gas in natural fractures and matrix pore structure and adsorbed gas in organic materials. Langmuir isotherm is widely used to describe the gas adsorption phenomenon. The amount of gas stored in shale is often described by Langmuir equation:

$$G_s = \frac{V_L P}{P + P_L},$$

(4)

where G_s is the gas content in scf/ton, V_l is the Langmuir volume in scf/ton, P_l is the Langmuir pressure in psi, and P is pressure in psi.

The bulk density of shale (ρ_B) is needed to convert the typical gas content in to scf/ton. Langmuir pressure and Langmuir volume are two key parameters. Langmuir volume is referred to as the gas volume at the infinite pressure representing the maximum storage capacity for gas; Langmuir pressure is referred to as the pressure corresponding to one-half Langmuir volume. As the reservoir pressure is decreased, gas is desorbed from the surface of the matrix. Figure 2 shows a graph of gas content with pressure for the adsorbed gas and total gas used for Barnett Shale [4]. Both free gas and adsorbed gas add together to generate the total gas content. In Barnett Shale, the adsorbed gas is approximately 46% of the total gas. Contrary to conventional gas reservoirs, the amount of gas desorption in the matrix is commonly described by the Langmuir isotherm in a range of reservoir pressures. The Langmuir isotherm of the Barnett Shale used in this study is illustrated in Figure 3. It is clearly shown that higher Langmuir pressure releases more adsorbed gas and results in higher gas production. Generally, in early stage of production, when reservoir pressure is high, the gas desorption contribution to the gas production is insignificant; however, for long-term production, it is necessary to account for gas desorption, based on a laboratory measured isotherm due to the more substantial pressure depletion, resulting in more gas desorption. CMG [18] was used to model the effect of gas desorption from a shale gas reservoir in a black oil model with a technique developed by Seidle and Arri [19]. A Langmuir isotherm is replicated by a black oil model's solution gas ratio to include the effect of gas desorption in shale.

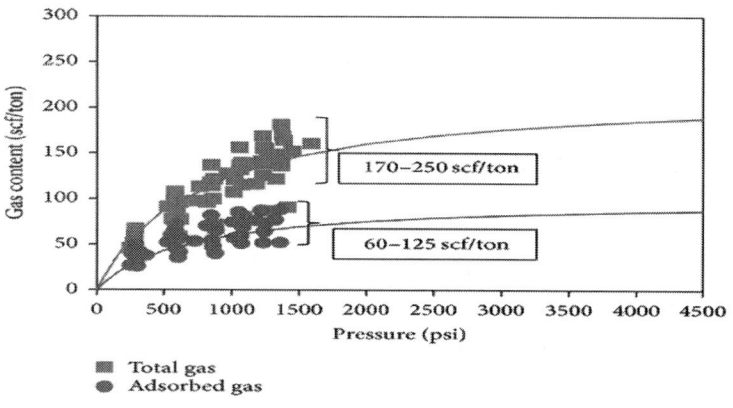

Figure 2: Adsorption isotherms for Barnett Shale core samples [4].

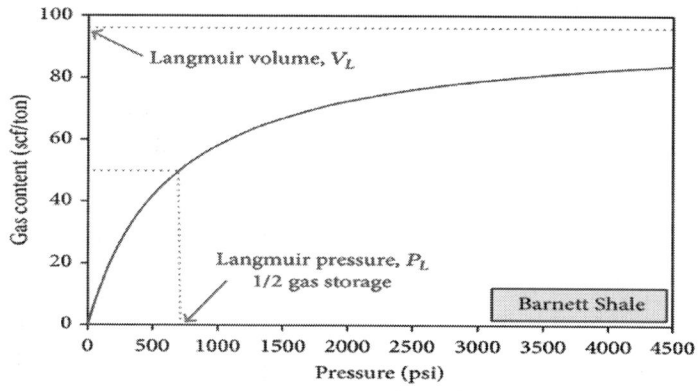

Figure 3: Langmuir isotherm curve for Barnett Shale (V_I=96 scf/ton, P_I=650 psi, and ρ_B =2058 g/cm³).

HISTORY MATCHING FOR BARNETT SHALE

Published average reservoir data for a Barnett Shale well were used for history matching [20]. In this case, the well was stimulated by a multistage fracturing with a single, perforated interval for each stage. In this simulation study, we set up a reservoir with a volume of 3000ft×1500ft×300ft. The fracture spacing and half-length are set at 100 ft and 150 ft, respectively, and the number of fractures is 28. Detailed reservoir information about this section of the Barnett Shale is listed in Table 2. The reservoir is assumed to be homogeneous and the fractures evenly spaced, with stress-independent porosity and permeability. Only gas is flowing in the reservoir, which is assumed to behave as non-Darcy flow. The history matching of field data is presented in Figure 4(a). It shows a more reasonable match between the numerical simulation results and the actual field gas flow data, considering the effect of gas desorption, contributing to 15.6% of total gas production at around 4.5 years of gas production. In addition, Figure 4(b) shows the forecasting of gas production for a 30-year period with and without considering gas desorption. As shown, with the gas production, the gas desorption contributes more due to substantial pressure depletion and larger gas drainage area and finally contributes to 20.7% of the total gas production at 30 years of gas production. Thus, the impact of gas desorption cannot be ignored when performing history matching and assessing production forecast of gas production in Barnett Shale formation. Hence, this study takes into account gas desorption effect for the subsequent optimization of multiwell placement in Barnett Shale.

Table 2: Parameters used in history matching

Parameter	Value(s)	Unit
The model dimensions	3,000 (length) × 1,500 (width) × 300 (height)	ft
Initial reservoir pressure	2950	psi

Bottom hole pressure (BHP)	500	psi
Production time	30	year
Reservoir temperature	150	°F
Gas viscosity	0.0201	cp
Initial gas saturation	0.70	fraction
Total compressibility	3×10-6	Psi−1
Matrix permeability	0.00015	md
Matrix porosity	0.06	fraction
Fracture conductivity	1	md-ft
Fracture half-length	155	ft
Fracture spacing	100	ft
Fracture height	300	ft
Horizontal well length	2968	ft
Number of fractures	28	number

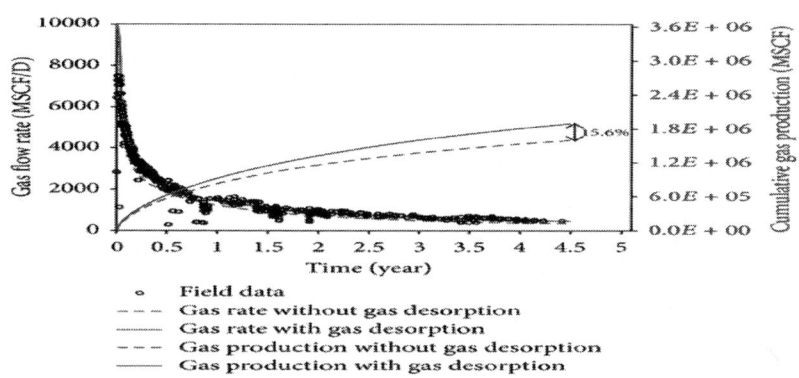

(a)

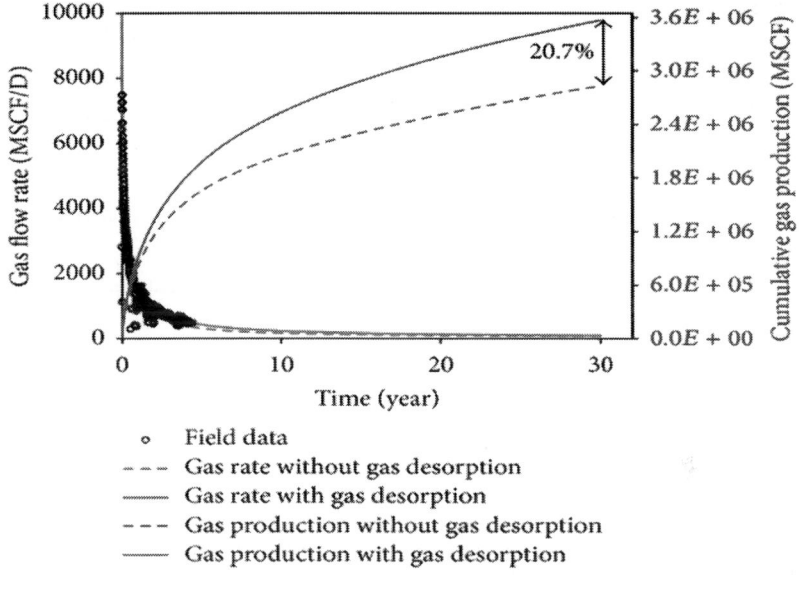

o   Field data
- - -   Gas rate without gas desorption
———   Gas rate with gas desorption
- - -   Gas production without gas desorption
———   Gas production with gas desorption

(b)

Figure 4: History matching of Barnett Shale with and without gas desorption effect.

MULTIWELL MODELING

Two scenarios describing multiple horizontal well placement were studied, as illustrated in Figure 5. Scenario 1 is referred to as aligning fracturing, where hydraulic fracturing is between two wells in an aligned pattern, and Scenario 2 is referred to as alternating fracturing, where hydraulic fracturing is between two wells in a staggered pattern. We investigate the effect of fracture spacing towards comparison of gas production between these two scenarios. We set up a shale gas reservoir model with a volume of 5000ft×1600ft×200ft. Detailed reservoir information used for Barnett Shale is listed in Table 3. Comparison of cumulative gas production is shown in Figure 6. It shows that there is almost no difference of gas production for these two scenarios when fracture spacing is below 400 ft; however, Scenario 2 yields higher gas production than Scenario 1 when the fracture spacing is 400 ft or above. Figure 7 presents the comparison of average reservoir pressure with fracture spacing of 600 ft. It can be seen that for Scenario 2, it has a larger average reservoir pressure drawdown, leading to higher cumulative gas production. Pressure distribution at 30 years of gas production for these two scenarios is shown in Figure 8. It shows that more contact reservoir area is drained effectively by Scenario 2, compared to Scenario 1. In addition, Rafiee et al. [7] reported that Scenario 2 design can increase the stress interference between fractures and create more effective stimulated reservoir volume to improve gas production. Therefore, Scenario 2 is used for optimization of multi-well placement in the subsequent study of this paper.

Table 3: Parameters used in multiwell modeling

Parameter	Value(s)	Unit
The model dimensions	5,000 (length) × 1,600 (width) × 200 (height)	ft
Initial reservoir pressure	3800	psi
BHP	500	psi
Production time	30	year
Reservoir temperature	180	°F
Gas viscosity	0.0201	cp
Initial gas saturation	0.70	fraction
Total compressibility	3×10^{-6}	Psi-1
Matrix permeability	0.0001	md
Matrix porosity	0.06	fraction
Fracture conductivity	50	md-ft
Fracture half-length	300	ft
Fracture spacing	200, 400, 600	ft
Fracture height	200	ft
Horizontal well length per well	3600	ft
Number of fractures per well	18, 9, 6	number
Number of wells	2	number
Well distance	620	ft

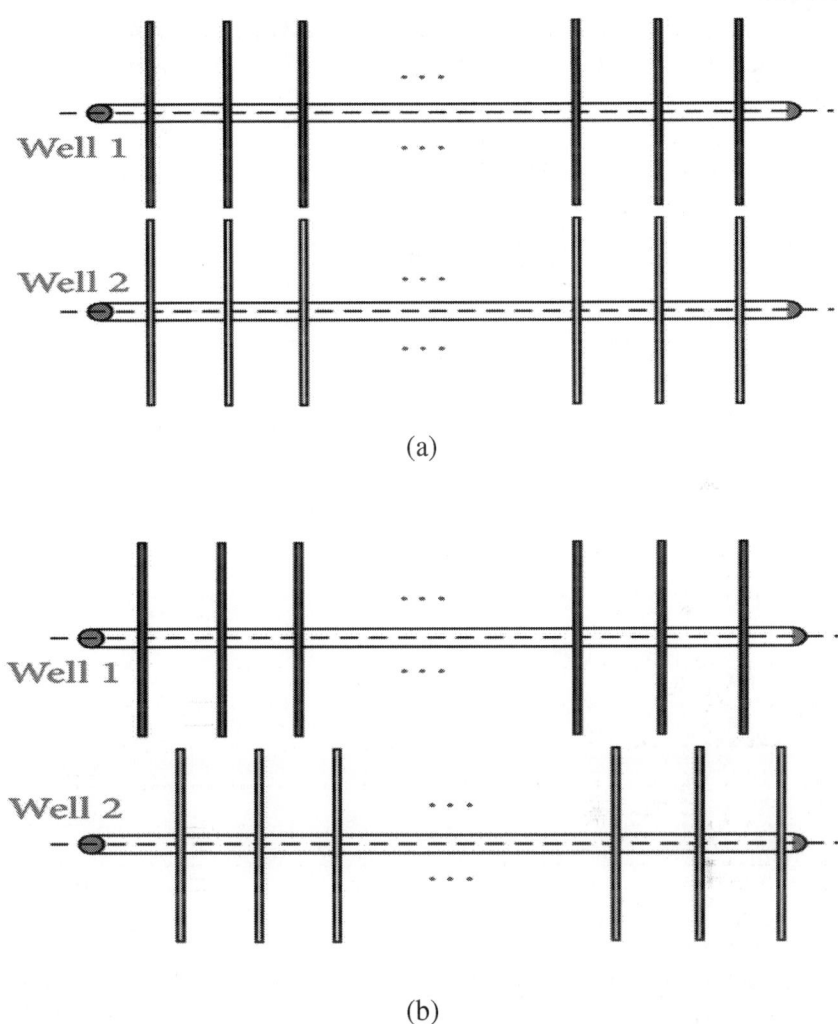

(a)

(b)

Figure 5: Two scenarios of multiple horizontal well placement.

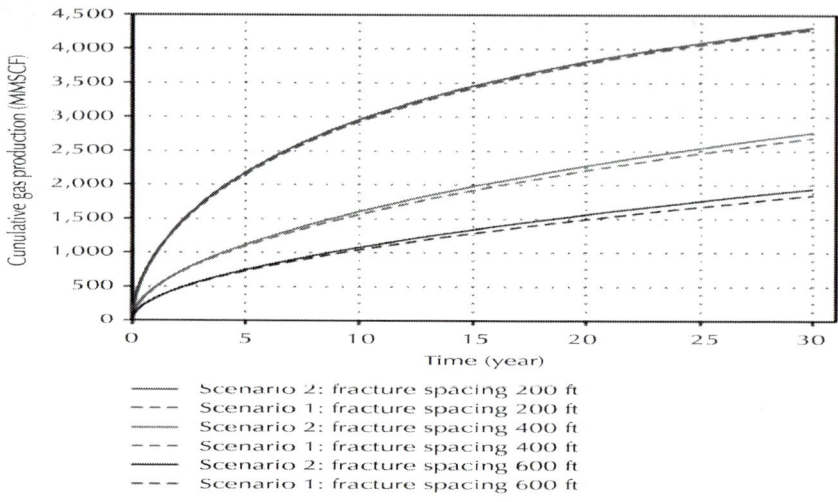

Figure 6: Comparison of cumulative gas production.

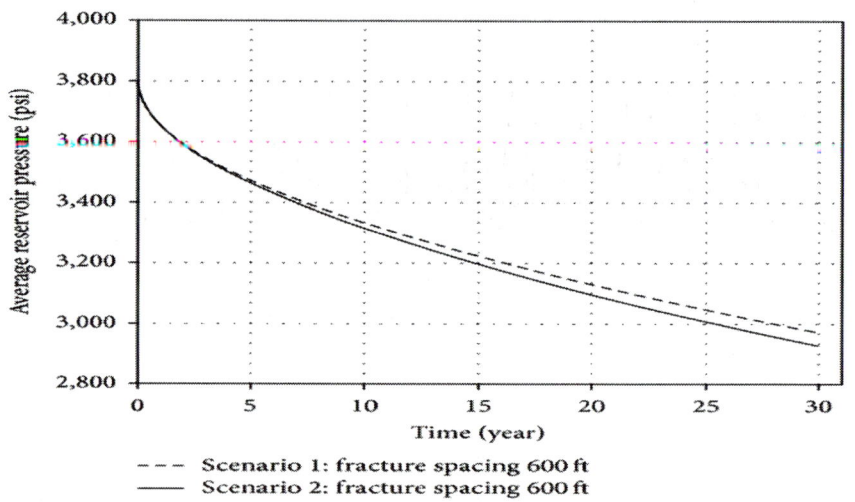

Figure 7: Comparison of average reservoir pressure.

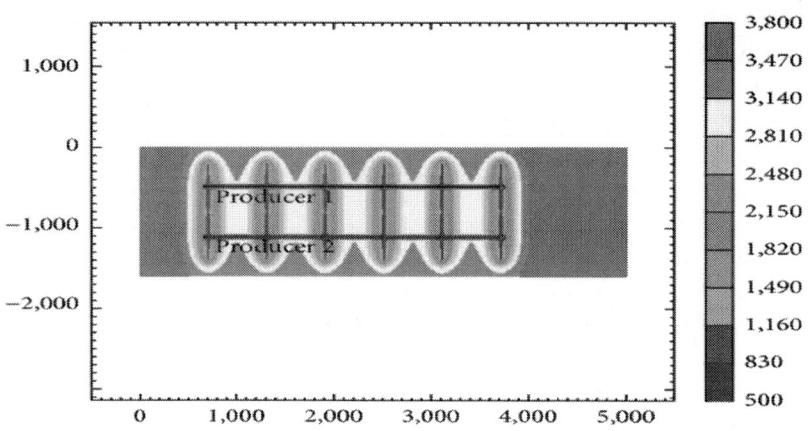

(a)

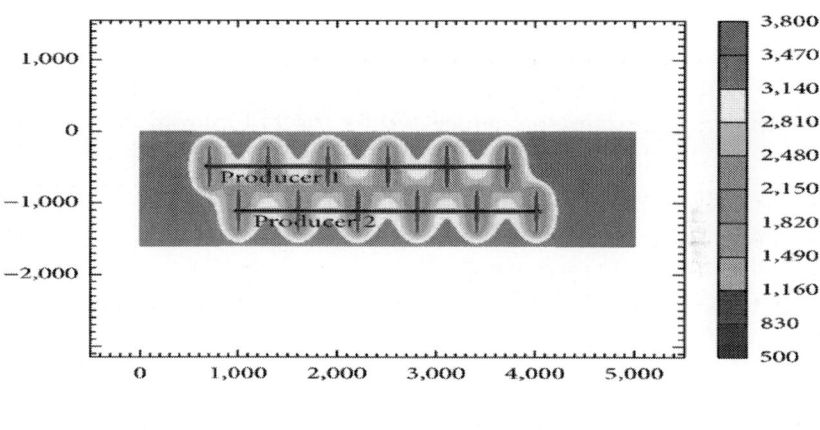

(b)

Figure 8: Pressure distribution at 30 years of gas production.

MULTIWELL OPTIMIZATION

Response surface methodology (RSM) approach is applied to the optimization of two horizontal well placement based on the Barnett

Shale reservoir information. RSM is utilized to approximate a response, in terms of maximum NPV in this work, over a region of interest specified by the range of variability of input factors based on the least squares criterion. The RSM model can be linear or fully quadratic. It can offer a cost-effective and efficient way to manage the uncertainties for shale gas reservoir development. More detailed mathematical and statistical theories of RSM can be found in the work by Myers and Montgomery [21]. We set up a shale gas reservoir model with a volume of 5000ft×2000ft×200ft. Detailed reservoir information used for the Barnett Shale is listed in Table 4. Six uncertain parameters such as porosity (A), permeability (B), fracture half-length (C), fracture conductivity (D), fracture spacing (E), and well distance (F) are given a reasonable range with the actual maximum and minimum values or coded symbol of "+1" and "−1," respectively, as shown in Table 5. The number of hydraulic fractures is 87 and 35, corresponding to fracture spacing of 40 ft and 100 ft, respectively. These uncertainty ranges come from field data, analogues, and history matching of the Barnett Shale reservoirs. According to 6 variables, 38 cases were required, based on the approach of D-Optimal design, which was originated from the optimal design theory [22]. Table 6 shows the 38 combinations of these uncertain parameters generated by the D-Optimal design. After numerical simulation of each case, cumulative gas production and gas flow rate are obtained and shown in Figures 9 and 10 for selected cases with various fracture spacing and fracture half-length. It clearly shows that the cumulative gas production at 30 years of gas production is in the range of 3934–6529 MMSCF and gas flow rate has a large uncertainty. This means further optimization is needed.

Table 4: Parameters used in multiwell optimization

Parameter	Value(s)	Unit
The model dimensions	5,000 (length) × 2,000 (width) × 200 (height)	ft
Initial reservoir pressure	3800	psi
BHP	500	psi
Production time	30	year
Reservoir temperature	180	°F
Gas viscosity	0.0201	cp
Initial gas saturation	0.70	fraction
Total compressibility	3×10-6	Psi−1
Fracture height	200	ft
Horizontal well length per well	3500	ft
Number of wells	2	number

Table 5: Uncertainty parameters in this study

Parameters	Coded symbol	Minimum (−1)	Maximum (+1)	Unit
Porosity	A	0.04	0.08	fraction
Permeability	B	0.00005	0.0005	md
Fracture half-length	C	200	400	ft
Fracture conductivity	D	1	50	md-ft
Fracture spacing	E	40	100	ft
Well distance	F	500	1000	ft

Table 6: D-Optimal design table

Run	A—Porosity	B—Permeability (md)	C—Fracture half-length (ft)	D—Fracture conductivity (md-ft)	E—Fracture spacing (ft)	F—Well distance (ft)
1	0.05	5.00E-05	200	1	40	900
2	0.04	5.00E-04	400	41	100	900
3	0.06	2.52E-04	300	1	60	500
4	0.08	3.38E-04	300	27	40	700
5	0.06	5.00E-04	300	25	80	900
6	0.05	2.37E-04	200	50	40	600
7	0.08	5.00E-05	400	1	40	700
8	0.08	3.38E-04	300	27	40	700
9	0.05	5.00E-04	200	3	100	500
10	0.05	7.25E-05	400	12	100	700
11	0.04	4.71E-04	200	28	40	1000
12	0.04	4.53E-04	400	41	80	500
13	0.08	5.00E-05	400	50	60	600
14	0.08	5.00E-04	200	50	60	500
15	0.04	3.76E-04	300	1	100	1000
16	0.08	3.20E-04	400	50	100	1000
17	0.06	5.00E-04	300	25	80	800
18	0.04	5.00E-05	300	50	60	900
19	0.04	5.00E-04	200	50	100	900
20	0.08	2.71E-04	200	15	80	800
21	0.06	5.00E-05	200	33	100	1000
22	0.04	5.00E-05	400	26	40	500
23	0.04	5.00E-04	300	9	60	1000
24	0.08	5.00E-04	200	1	40	1000
25	0.06	5.00E-05	400	43	40	900
26	0.07	1.92E-04	300	45	100	500
27	0.06	2.52E-04	300	1	60	500
28	0.08	2.71E-04	200	15	80	800
29	0.04	5.00E-05	200	17	80	500
30	0.08	5.00E-05	200	14	40	500
31	0.08	5.00E-05	300	1	100	1000
32	0.04	5.00E-04	400	1	40	700
33	0.08	5.00E-04	400	1	100	500
34	0.06	5.00E-04	300	25	80	800
35	0.08	5.00E-05	200	50	40	1000
36	0.08	5.00E-04	400	50	80	600
37	0.06	5.00E-04	400	50	40	1000
38	0.07	1.98E-04	400	14	60	1000

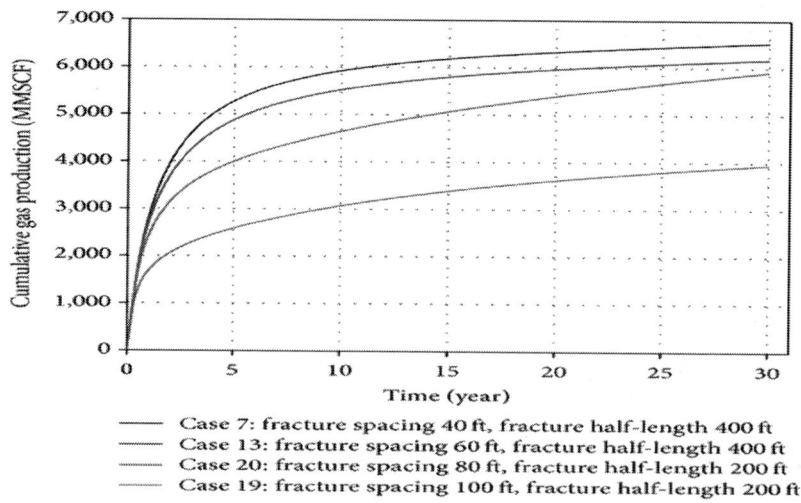

Case 7: fracture spacing 40 ft, fracture half-length 400 ft
Case 13: fracture spacing 60 ft, fracture half-length 400 ft
Case 20: fracture spacing 80 ft, fracture half-length 200 ft
Case 19: fracture spacing 100 ft, fracture half-length 200 ft

Figure 9: Cumulative gas production of 30 run cases for a 30-year period.

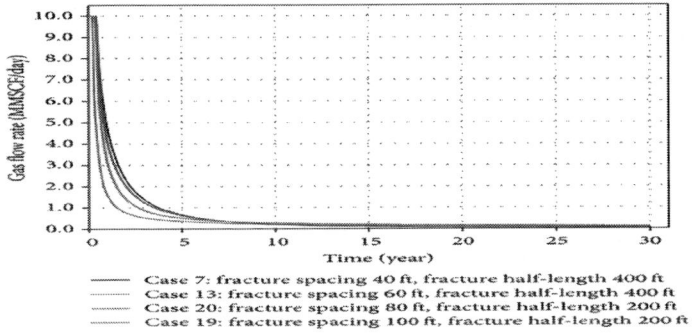

Figure 10: Gas flow rate of 30 run cases for a 30-year period.

Once the cumulative gas production of 38 run cases obtained, the economic Excel spreadsheet is used to calculate the corresponding NPVs. In this work, the impact of different gas price is considered for calculation of NPVs and optimization of fracturing design in multi-well placement to determine the optimal stimulation design. The price of natural gas plays a critical role in economic success of shale gas development. The activity in shale gas exploitation increases with the increasing gas price. Negative NPV of the project suggests that an unoptimized shale gas production can result in not only no income but also loss of money. For gas price of $3/MSCF, $4/MSCF, and $5/MSCF, the positive NPV is in the range of $0.36–13.39 million, $2.10–18.88 million, and $3.42–24.38 million, respectively, as shown in Figure 11.

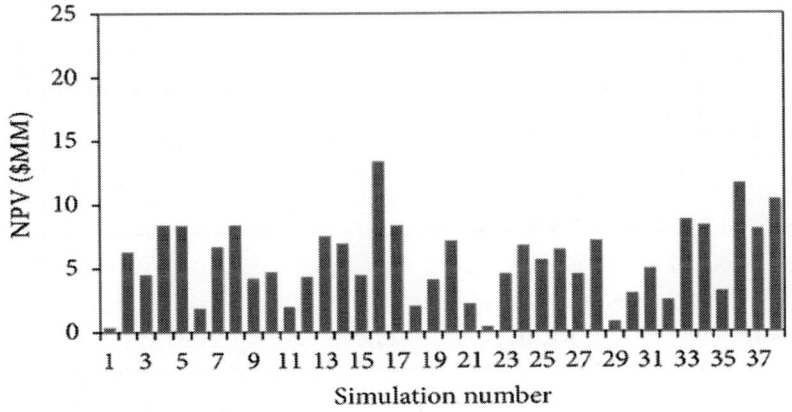

(a)

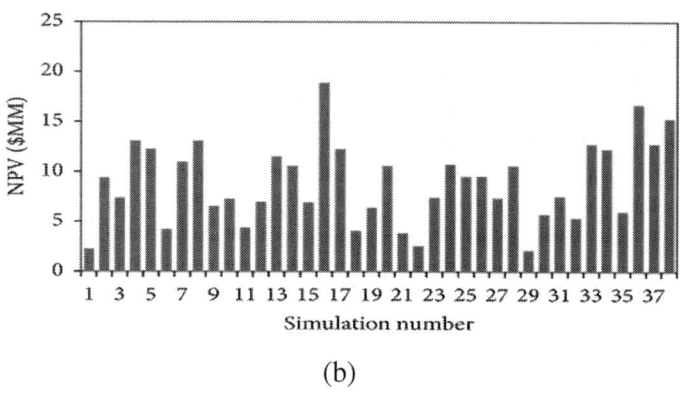

(b)

(c)

Figure 11: NPVs of 38 simulation cases at 30 years of gas production with different gas prices.

Once NPVs of 38 run cases obtained, the Design-Expert Software is used to build the NPV response surface model in this study. To select the appropriate model, the statistical approach was used to determine which polynomial fits the equation among linear model, two-factor interaction model (2FI), quadratic model, and cubic model, as shown in Tables 7, 8, and 9, for different gas prices. The criterion for selecting the appropriate model is choosing the highest polynomial model, where the additional terms are significant and the model is not aliased. Although the cubic model is the highest polynomial model, it is not selected because it is aliased. Aliasing is a result of reducing the number of experimental runs. When it occurs, several groups of effects are combined into one group and the most significant effect in the group is used to represent the effect of the group. Essentially, it is important that the model is not aliased. In addition, other criteria are to select the model that has the maximum "Adjusted R-Squared" and "Predicted R-Squared". Thus, the fully quadratic model is selected to build the NPV response surface in the subsequent optimization process.

Table 7: Statistical approach to select the RSM model with gas price of $3/MSCF

Source	Std. Dev.	R-Squared	Adjusted R-Squared	Predicted R-Squared	Press	
Linear	0.27	0.8812	0.8582	0.8094	3.72	—
2FI	0.26	0.9433	0.8688	0.2155	15.31	—
Quadratic	0.064	0.9979	0.9921	0.9032	1.89	Suggested
Cubic	0	1	1	—	—	Aliased

Table 8: Statistical approach to select the RSM model with gas price of $4/MSCF

Source	Std. Dev.	R-Squared	Adjusted R-Squared	Predicted R-Squared	Press	
Linear	0.25	0.8950	0.8747	0.8314	3.16	—
2FI	0.24	0.9497	0.8837	0.2936	13.26	—
Quadratic	0.065	0.9978	0.9917	0.8851	2.16	Suggested
Cubic	0	1	1	—	—	Aliased

Table 9: Statistical approach to select the RSM model with gas price of $5/MSCF

Source	Std. Dev.	R-Squared	Adjusted R-Squared	Predicted R-Squared	Press	
Linear	0.26	0.8985	0.8788	0.8368	3.37	—
2FI	0.25	0.9525	0.8901	0.3213	14.01	—
Quadratic	0.071	0.9975	0.9908	0.8712	2.66	Suggested
Cubic	0	1	1	—	—	Aliased

The equation fitted to the NPV response surface with the coded symbol for different gas prices is presented below.For gas price of $3/MSCF,

$$\sqrt{\text{NPV}} = 2.67 + 0.59 \times A + 0.35 \times B + 0.35 \times C$$
$$+ 0.056 \times D + 0.21 \times E + 0.170 \times F - 0.025 \times AB$$
$$+ 0.058 \times AC - 0.011 \times AD - 0.160 \times AE$$
$$- 0.037 \times AF - 0.090 \times BC - 0.026 \times BD$$
$$- 0.028 \times BE - 0.0056 \times BF + 0.15 \times CD$$
$$+ 0.0098 \times CE + 0.13 \times CF + 0.046 \times DE$$
$$+ 0.044 \times DF - 0.064 \times EF - 0.086 \times A^2$$

$$- 0.24 \times B^2 - 0.013 \times C^2 - 0.069 \times D^2$$

$$- 0.25 \times E^2 - 0.057 \times F^2.$$

(5)

For gas price of $4/MSCF,

$$\sqrt{\text{NPV}} = 3.27 + 0.60 \times A + 0.35 \times B + 0.38 \times C$$
$$+ 0.050 \times D + 0.12 \times E + 0.190 \times F + 0.014 \times AB$$
$$+ 0.055 \times AC + 0.018 \times AD - 0.11 \times AE$$
$$- 0.018 \times AF - 0.088 \times BC - 0.0012 \times BD$$
$$+ 0.028 \times BE + 0.011 \times BF + 0.020 \times CD$$
$$+ 0.0014 \times CE + 0.12 \times CF + 0.055 \times DE$$
$$+ 0.027 \times DF - 0.062 \times EF - 0.082 \times A^2$$
$$- 0.25 \times B^2 - 0.0083 \times C^2 - 0.094 \times D^2$$
$$- 0.19 \times E^2 - 0.062 \times F^2.$$

(6)

For gas price of $5/MSCF,

$$\sqrt{NPV} = 3.78 + 0.63 \times A + 0.37 \times B + 0.410 \times C$$
$$+ 0.053 \times D + 0.070 \times E + 0.20 \times F$$
$$+ 0.027 \times AB + 0.060 \times AC + 0.027 \times AD$$
$$- 0.096 \times AE - 0.011 \times AF - 0.091 \times BC$$
$$+ 0.0042 \times BD + 0.046 \times BE + 0.017 \times BF$$

$$+ 0.021 \times CD - 0.00061 \times CE + 0.13 \times CF$$
$$+ 0.062 \times DE + 0.024 \times DF - 0.063 \times EF$$
$$- 0.083 \times A^2 - 0.26 \times B^2 - 0.0052 \times C^2$$
$$- 0.11 \times D^2 - 0.17 \times E^2 - 0.07 \times F^2,$$

$$(7)$$

where A is formation porosity; B is formation permeability, md; C is fracture half-length, ft; D is fracture conductivity, md-ft; E is fracture spacing, ft; F is well distance, ft.

The normal plot of residuals, reflecting the distribution of the residuals, for different gas prices is shown in Figure 12. All the points in the "Normal Plot of Residuals" fall on the straight line, meaning the residuals are normally distributed. Figure 13 shows the plot of "Predicted versus Actual" for different gas prices, illustrating whether the generated equation of NPV response surface accurately predicts the actual NPV values. It can be seen that generated NPV response surface models provide such reliable predicted values of NPV, as compared with the actual values of NPV. This means that the generated NPV response surface models are reliable.

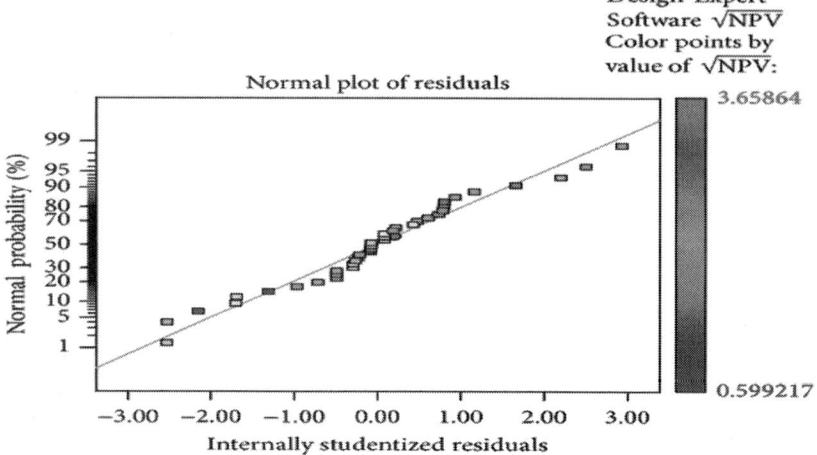

(a)

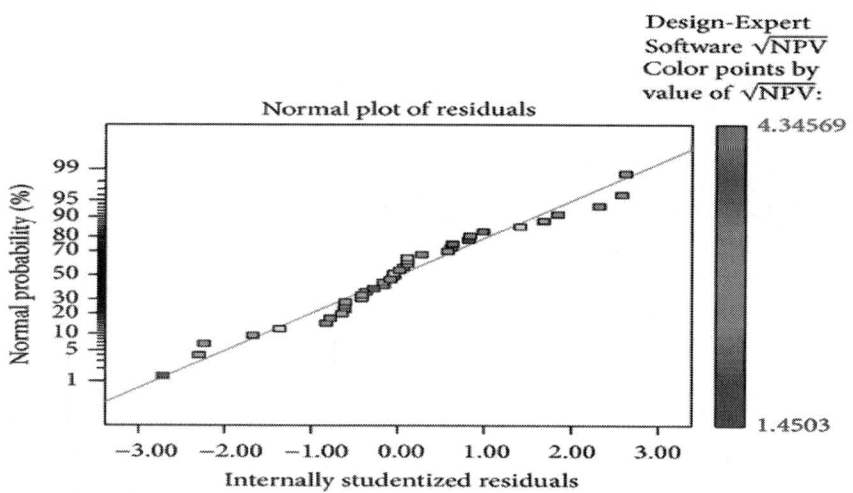

(b)

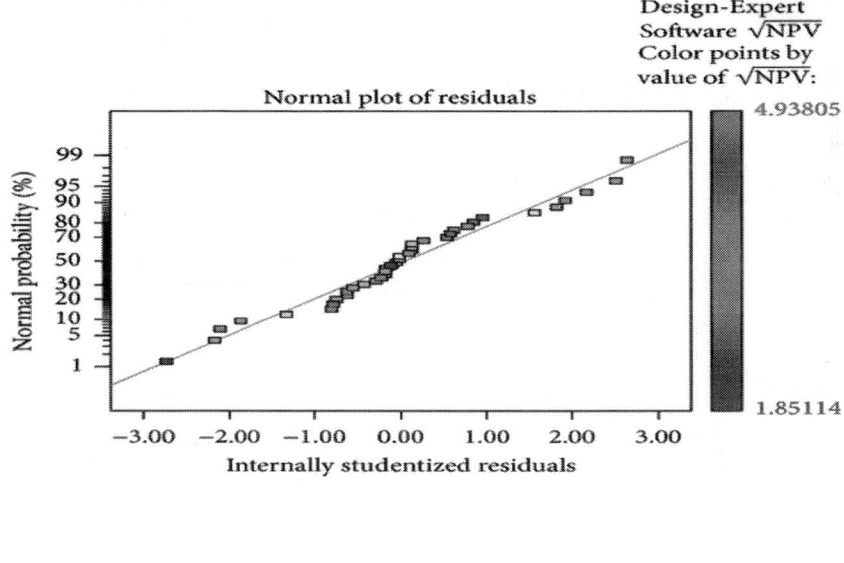

(c)

Figure 12: Normal plot of residuals.

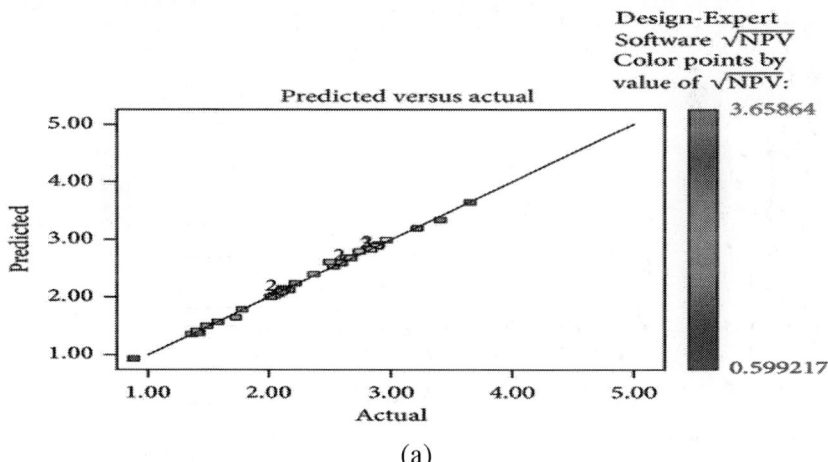

(a)

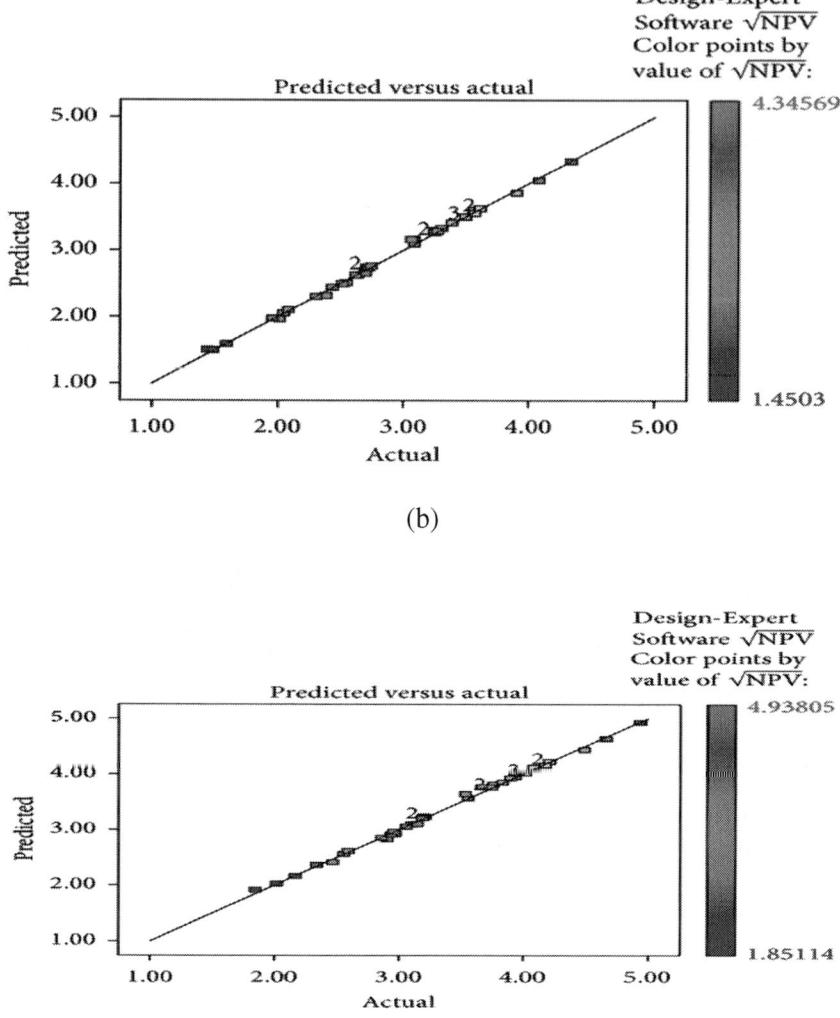

Figure 13: Predicted NPV versus the actual NPV plot.

Figure 14 shows the 3D surface of the well distance and fracture spacing with gas price of $4/MSCF. It shows that there exists an optimal combination between well distance and fracture spacing. For fracture half-length of 200 ft, the NPV first increases and then decrease with increasing well distance. For fracture half-length of 400 ft, the NPV increases with increasing well distance within the range of this study. With the increasing fracture half-length, the optimal point moves to larger well distance. Figure 15 presents the 3D surface of the well distance and fracture half-length with gas price of $4/MSCF. It can be seen that larger well distance and fracture half-length will lead to higher NPV. Similarly, Figure 16 presents the 3D surface of the fracture spacing and fracture half-length with gas price of $4/MSCF. It shows that there exists an optimal value for fracture spacing. With the increasing well distance, the optimal point moves to larger fracture half-length and smaller fracture spacing. Therefore, it can provide some insights into optimization of stimulation designs and completion strategies to obtain the maximum economic viability of the field.

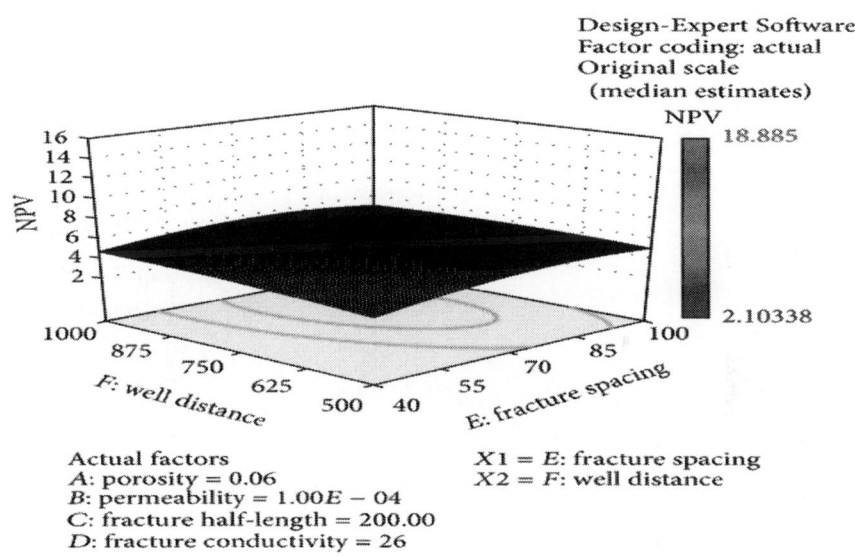

(a)

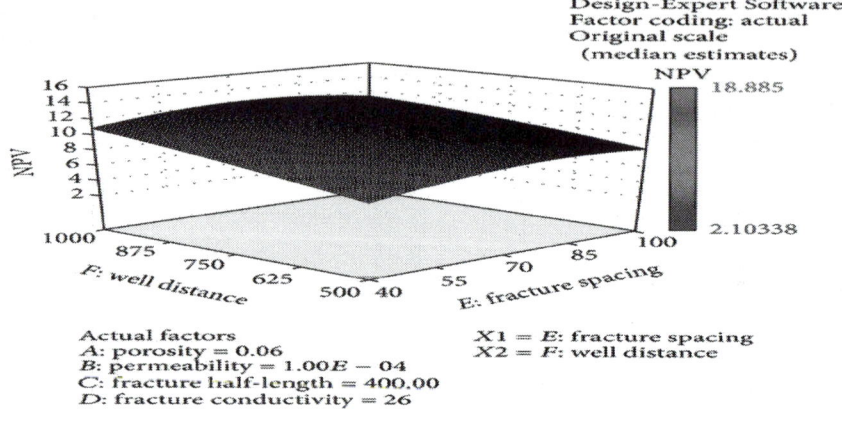

(b)

Figure 14: 3D surface of NPV at varied values of well distance and fracture spacing with gas price of $4/MSCF.

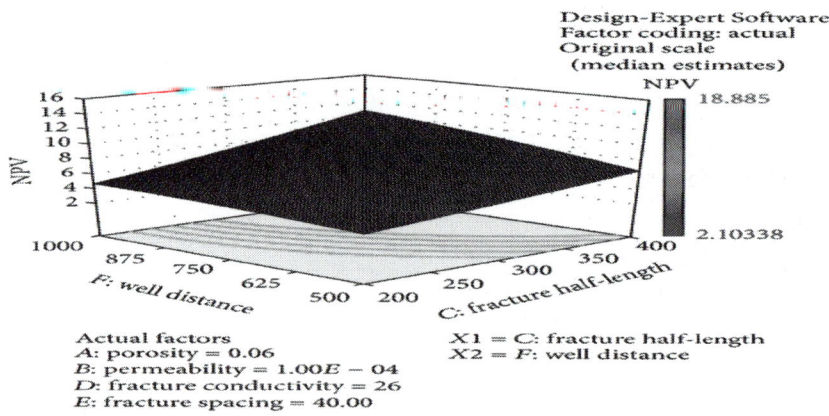

(a)

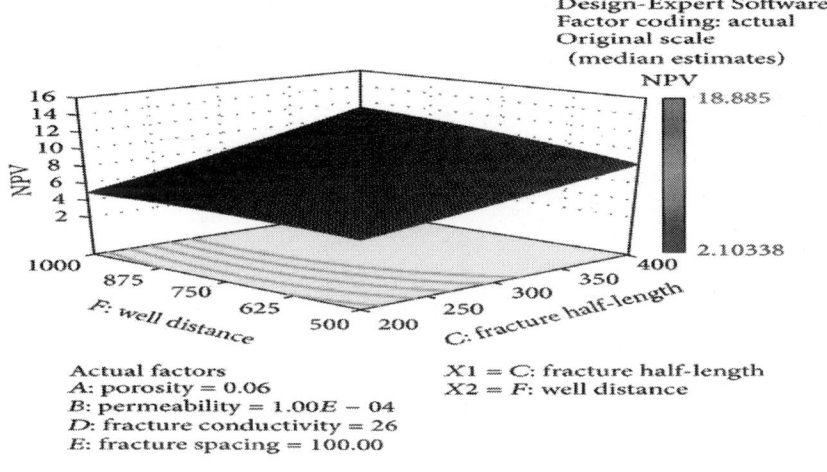

(b)

Figure 15: 3D surface of NPV at varied values of well distance and fracture half-length with gas price of $4/MSCF.

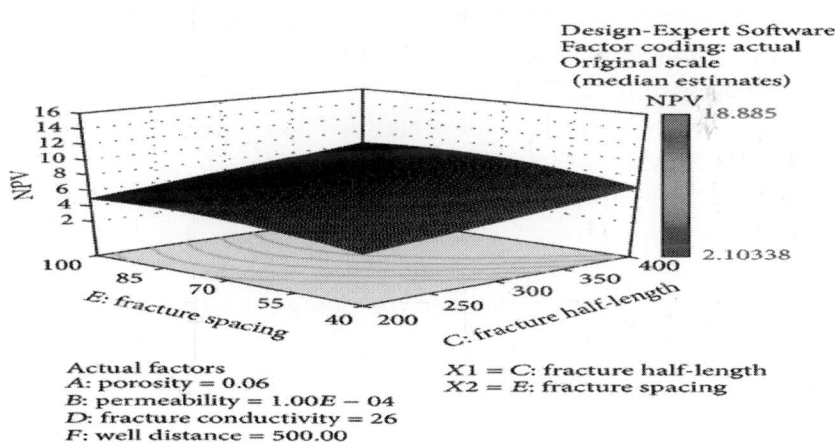

(a)

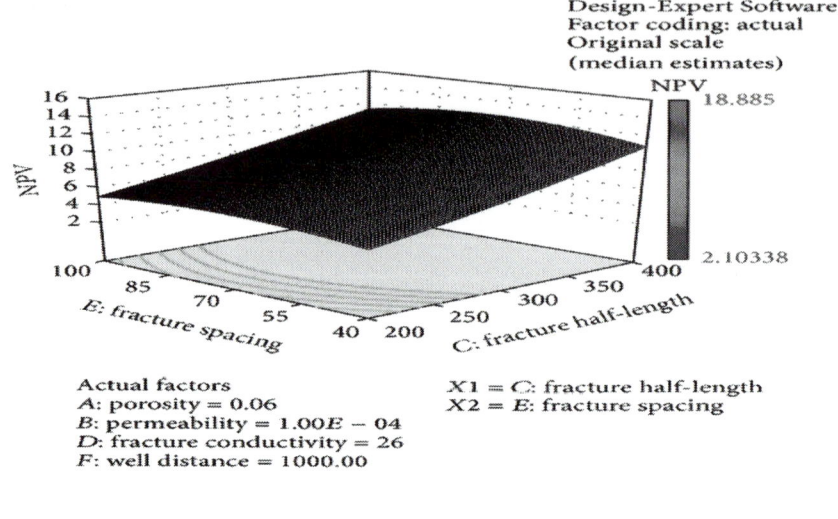

(b)

Figure 16: 3D surface of NPV at varied values of fracture spacing and fracture half-length with gas price of $4/MSCF.

The numerical optimization option selects the set of variables that leads to the maximum NPV value. In this study for the Barnett Shale, the optimal designs with porosity of 0.06 and permeability of 0.0001 md and different gas prices are listed in Table 10. The optimal NPV value is $8.70 MM, $12.40 MM, and $16.34 MM, for gas price of $3/MSCF, $4/MSCF, and $5/MSCF, respectively. What is more, the optimal fracture spacing is 80 ft for gas price of $3/MSCF, 70 ft for gas price of $4/MSCF, and 60 ft for gas price of $5/MSCF, as shown in Figure 17. It is extremely vital to validate the optimal results. Verification is performed by running the simulation with the best design condition. Figure 18 shows the cumulative gas production and gas flow rate for these three optimum cases with different gas prices, respectively. The NPV is calculated as $7.91 MM, $11.81 MM, and $15.75 MM, for gas price of $3/MSCF, $4/MSCF, and $5/MSCF, respectively. The absolute NPV difference between the NPV from the response surface and the real NPV is small. As indicated by the small relative error, all three solutions show a very good agreement between the calculated NPV and the real NPV.

Table 10: Optimal combinations and optimization validation

Porosity	Permeability (md)	Fracture half-length (ft)	Fracture conductivity (md-ft)	Fracture spacing (ft)	Well distance (ft)	Gas price ($/MSCF)	Optimal NPV ($MM)	Validated NPV ($MM)	Relative error
0.06	0.0001	400	26	80	1000	3	8.70	7.91	0.091
0.06	0.0001	400	26	70	1000	4	12.40	11.81	0.047
0.06	0.0001	400	26	60	1000	5	16.34	15.75	0.036

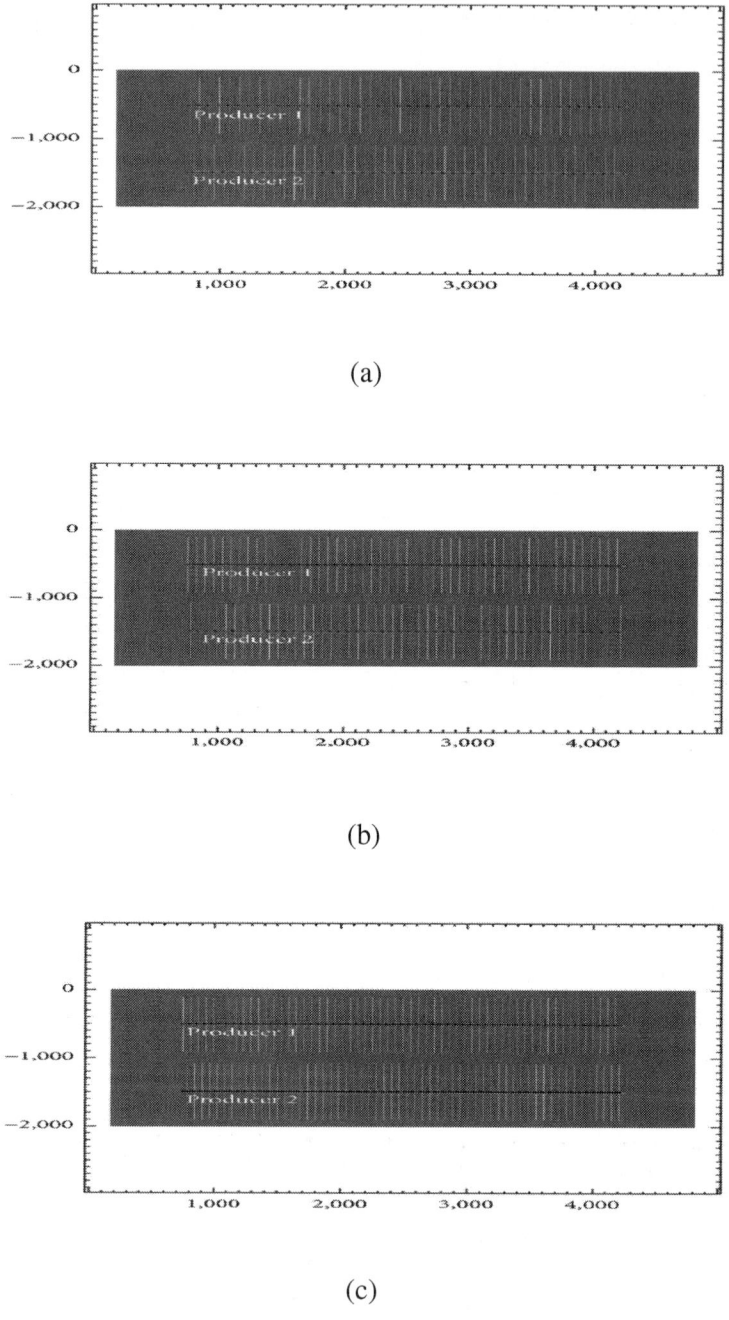

Figure 17: The optimum fracture spacing with different gas prices.

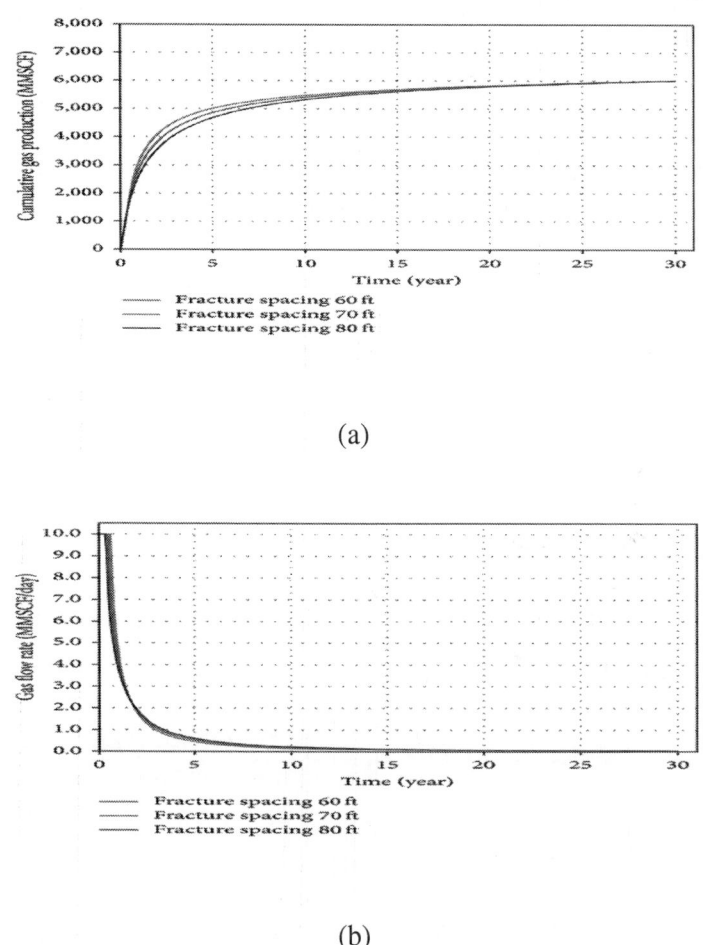

(a)

(b)

Figure 18: Three optimum cases with different fracture spacing for different gas prices.

SUMMARY AND CONCLUSIONS

The economic success of shale gas reservoirs depends on optimization of the number of treatment stages and number of fractures and horizontal wells. In this paper, response surface methodology was used to obtain the optimal design for shale gas production by optimizing reservoir and fracture uncertainty parameters in two horizontal wells.

We applied this method to optimize 6 uncertain parameters, such as permeability, porosity, fracture spacing, fracture half-length, fracture conductivity, and well distance for the Barnett Shale development. Also, the gas desorption effect is considered for modeling shale gas production because of its large contribution to the estimated ultimate recovery for the Barnett Shale. The impact of varying natural gas price is taken into account during the optimization process. The following conclusions can be drawn from this study.

- With a porosity of 0.06, a permeability of 0.0001 md, and a fracture conductivity of 26 md-ft, the optimal design combinations for Barnett Shale is fracture half-length of 400 ft, well distance of 1000 ft, and fracture spacing of 80 ft, 70 ft, and 60 ft for gas price of $3/MSCF, $4/MSCF, and $5/MSCF, respectively.

- The gas desorption contributes to 20.7% of EUR at 30 years of gas production for an actual Barnett Shale horizontal well.

- The proposed approach is practical and efficient for the design and optimization of hydraulic fracturing for multiple horizontal wells in shale gas reservoirs.

ACKNOWLEDGMENTS

Funding for this project is provided by the Hilcorp Energy Company. The authors would like to acknowledge Computer Modeling Group Ltd. for providing with usage of CMG-IMEX software. The authors would also like to express their gratitude for reviewers for their careful review of this paper.

REFERENCES

1. J. L. Miskimins, "Design and life-cycle considerations for unconventional-reservoir wells," SPE Production and Operations, vol. 24, no. 2, pp. 353–359, 2009.

2. G. Waters, B. Dean, R. Downie, K. Kerrihard, L. Austbo, and B. McPherson, "Simultaneous hydraulic fracturing of adjacent horizontal wells in the woodford shale," in Proceedings of the SPE Hydraulic Fracturing Technology Conference, pp. 694–715, Woodlands, Tex, USA, January 2009.

3. M. J. Kaiser, "Haynesville shale play economic analysis," Journal of Petroleum Science and Engineering, vol. 82-83, pp. 75–89, 2012. · View at Google Scholar ·

4. S. L. Montgomery, D. M. Jarvie, K. A. Bowker, and R. M. Pollastro, "Mississippian Barnett Shale, Fort Worth basin, north-central Texas: gas-shale play with multi-trillion cubic foot potential," AAPG Bulletin, vol. 89, no. 2, pp. 155–175, 2005. · ·

5. M. Segatto and I. Colombo, "Use of reservoir simulation to help gas shale reservoir estimation," inProceedings of the International Petroleum Technology Conference (IPTC ‹11), Bangkok, Thailand, 2011.

6. S. Esmaili, A. Kalantari-Dahaghi, and S. D. Mohaghegh, "Modeling and history matching of hydrocarbon production from Marcellus shale using data mining and pattern recognition technologies," in Proceedings of the SPE Eastern Regional Meeting (SPE ‹12), Lexington, Ky, USA, 2012.

7. M. Rafiee, M. Y. Soliman, and E. Pirayesh, "Hydraulic fracturing design and optimization: a modification to zipper frac," in Proceedings of the SPE Eastern Regional Meeting (SPE ‹12), Lexington, Ky, USA, 2012.

8. O. C. Díaz de Souza, A. J. Sharp, R. C. Martinez et al., "Integrated unconventional shale gas reservoir modeling: a worked example from the Haynesville Shale, De Soto Parish, North Lousiana," inProceedings of the Americas Unconventional Resources Conference (SPE ‹12), Pittsburgh, Pa, USA, 2012.

9. J. Harpel, L. Barker, J. Fontenot, C. Carroll, S. Thomson, and K. Olson, "Case history of the Fayetteville shale completions," in Proceedings of the SPE Hydraulic Fracturing Technology Conference (SPE ‹12), The Woodlands, Tex, USA, 2012.

10. H. Ramakrishnan, R. Yuyan, and J. Belhadi, "Real-time completion optimization of multiple laterals in gas shale reservoirs: Integration of geology, log, surface seismic, and microseismic information," inProceedings of the SPE Hydraulic Fracturing Technology Conference, pp. 691–705, The Woodlands, Tex, USA, January 2011.

11. S. Tavassoli, W. Yu, F. Javadpour, and K. Sepehrnoori, "Well screen and optimal time of refracturing: a Barnett shale well," Journal of Petroleum Engineering, vol. 2013, pp. 1–10, 2013.

12.   W. Yu and K. Sepehrnoori, "Simulation of gas desorption and geomechanics effects for unconventional gas reservoirs," in Proceedings of the SPE Western Regional and AAPG Pacific Section Meeting (SPE ‹13), Monterey, Calif, USA,, 2013.

13.   W. Yu and K. Sepehrnoori, "An efficient reservoir simulation approach to design and optimize unconventional gas production," in Proceedings of the SPE Western Regional and AAPG Pacific Section Meeting (SPE ‹13), Monterey, Calif, USA,, 2013.

14.   B. Rubin, "Accurate simulation of non-darcy flow in stimulated fractured shale reservoirs," inProceedings of the SPE Western Regional Meeting (SPE ‹10), pp. 19–34, Anaheim, Calif, USA, May 2010.

15.   C. L. Cipolla, E. P. Lolon, J. C. Erdle, and B. Rubin, "Reservoir modeling in shale-gas reservoirs," SPE Reservoir Evaluation and Engineering, vol. 13, no. 4, pp. 638–653, 2010.

16.   R. D. Evans and F. Civan, "Characterization of non-Darcy multiphase flow in petroleum bearing formations," U.S. DOE Contract DE-AC22-90BC14659, School of Petroleum and Geological Engineering, University of Oklahoma, 1994.

17.   R. Schweitzer and H. I. Bilgesu, "The role of economics on well and fracture design completions of marcellus shale wells," in Proceedings of the SPE Eastern Regional Meeting (SPE ‹09), pp. 423–428, September 2009.

18.   CMG, IMEX User's Guide, Computer Modeling Group, 2011.

19.   J. P. Seidle and L. E. Arri, "Use of conventional reservoir models for coalbed methane simulation," inProceedings of the CIM/SPE International Technical Meeting (SPE ‹90), Calgary, Canada, 1990.

20.   H. A. Al-Ahmadi, S. Aramco, and R. A. Wattenbarger, "Triple-porosity models: one further step towards capturing fractured reservoirs heterogeneity," in Proceedings of the SPE/DGS Saudi Arabia Section Technical Symposium and Exhibition (SPE ‹11), Al-Khobar, Saudi Arabia, 2011.

21.   R. H. Myers and D. C. Montgomery, Response Surface Methodology: Process and Product Optimization Using Designed Experiments, John Wiley and Sons, Hoboken, NJ, USA, 2002.

22. J. Kiefer and J. Wolfowitz, "Optimum designs in regression problems," The Annals of Mathematical Statistics, vol. 30, no. 2, pp. 271–294, 1959.

Chapter 2

Compositional Modeling for Optimum Design of Water-Alternating CO_2-LPG EOR under Complicated Wettability Conditions

Jinhyung Cho, Sung Soo Park, Moon Sik Jeong, and Kun Sang Lee

Department of Natural Resources and Environmental Engineering, Hanyang University, 222 Wangsimni-ro, Seongdong-gu, Seoul 133-791, Republic of Korea

ABSTRACT

The addition of LPG to the CO_2 stream leads to minimum miscible pressure (MMP) reduction that causes more oil swelling and interfacial tension reduction compared to CO_2 EOR, resulting in improved oil recovery. Numerical study based on compositional simulation has been performed to examine the injectivity efficiency and transport behavior of water-alternating CO_2-LPG EOR. Based on oil, CO_2, and LPG prices, optimum LPG concentration and composition were designed for different wettability conditions. Results from this study indicate how injected LPG mole fraction and butane content in LPG affect lowering of interfacial tension. Interfacial tension reduction by supplement of LPG components leads to miscible condition causing more enhanced oil recovery. The maximum enhancement of oil recovery for oil-wet reservoir is 50% which is greater than 22% for water-wet reservoir According to the result of net present value (NPV) analysis at designated oil, CO_2, propane, and butane prices, the optimal injected LPG mole fraction and composition exist for maximum NPV. At the case of maximum NPV for oil-wet reservoir, the LPG fraction is about 25% in which compositions of propane and butane are 37% and 63%, respectively. For water-wet reservoir, the LPG fraction is 20% and compositions of propane and butane are 0% and 100%.

INTRODUCTION

CO_2 injection has been found to be an efficient method for oil recovery worldwide through a miscible or an immiscible displacement process. Mechanism of CO_2 enhanced oil recovery (EOR) is divided into two different processes, miscible flood and immiscible flood. Although miscible gas injection is a widely applied EOR process, it can be only applied when the reservoir pressure is higher than minimum miscible pressure (MMP). The main process of miscible gas injection is displacement efficiency improvement by oil viscosity reduction and swelling effect to reduce residual oil saturation. When reservoir pressure is higher than MMP, the injected CO_2 and reservoir oil are completely miscible and the displacement efficiency can be enhanced by zero interfacial tension [1]. Immiscible flood is usually applied when reservoir pressure is insufficient to miscible flood or reservoir oil contains many heavy components. The effects of immiscible flood are similar to miscible flood, but one major disadvantage is the limited solubility of CO_2 in oil, resulting in the restricted swelling effect and viscosity reduction.

Injected CO_2 and reservoir oil can be miscible by continuous contact. At the fore-end of injected fluid, CO_2 is persistently contacted with fresh oil following flow direction, and they are eventually miscible by the vaporizing-gas drive process. In contrast, at the back-end of injected

CO_2, near injection well, reservoir oil is continuously contacted with fresh CO_2 that causes the miscible state by the condensing-gas drive process [2]. CO_2 miscible flood making high enhanced oil recovery effect has a limit that it can be only applied when the reservoir pressure is higher than MMP. It can be settled by the application of CO_2-LPG EOR that is able to lower MMP less than that from the application of only CO_2 EOR. The addition of alkane solvents to the CO_2 injection generally accelerates swelling oil, reducing oil viscosity and decreasing the interfacial tension that can lead to better performance in enhancing oil recovery [3]. The effects of CO_2-LPG injection are verified by the experiment [4].

Figure 1 indicates the ternary diagram of phase behavior of reservoir oil and injected solvent. J and I signify injected fluid and reservoir oil. In the inner area of the ternary diagram curve, two phases of the reservoir fluid exist. In case of J_1-I_2, only CO_2 is injected into reservoir oil I_2. J_1 and I_2 cannot be miscible at the first contact because the line passes through the two-phase area. However, they arrive at miscible condition by multiple contact miscibility process. The J_1-I_1 line lies on the two-phase territory and both points J_1 and I_1 are located in the same side on the basis of limiting tie line. Therefore, J_1-I_1 cannot be miscible by first and multiple contact miscibility process. At J_2-I_1 and J_3-I_1 cases, first and miscible contact miscibility process is available. By the addition of LPG to the CO_2 stream, the location of injected solvent is moved from J_1 to J_2 or J_3 depending on the amount of injected LPG. It makes miscible condition from J_1-I_1 case that was not supposed to be miscible.

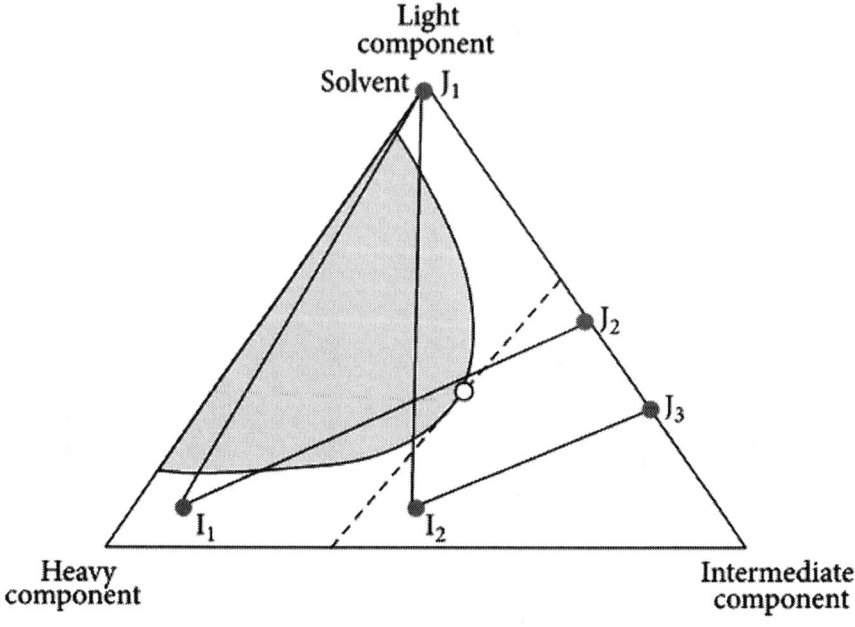

Figure 1: Phase ternary diagram for reservoir oil and injection gas relation [5].

To improve sweep efficiency, WAG (water-alternating-gas) process is applied to CO_2-LPG EOR method in this research. At the same WAG condition, injected LPG amount and composition are the variables considered in the study. Many experimental researches about the effects of LPG and impurities on MMP with oil have been actively developed [6]. Several established researches demonstrate the MMP reduction and oil recovery enhancement by CO_2-LPG EOR through only experimental ways [7, 8], but numerical approach to analyze the effectiveness of CO_2-LPG EOR was not included. Shokir [9] developed ACE algorithm model to analyze the effects of impurities on MMP between injected fluid and oil, but it could not explain how lower MMP affects oil recovery. Talbi et al. [3] conducted experimental research on oil swelling, viscosity reduction, interfacial tension reduction, and oil recovery improvement that resulted from injecting solvents into CO_2. However, if it is applied to field scale, it should be time consuming, so reservoir simulation model for CO_2-LPG flood is positively necessary. Recently, Teklu et al. [10] showed MMP reduction effect by CO_2-LPG flood in various reservoir scenarios using simulation model, but it focused on only the relationship between pore confinement, permeability, and MMP in shale reservoirs. It also does not make connection between MMP reduction and oil recovery in consideration of gas transport in porous media. Recent studies based on the modeling of spontaneous imbibition also indicated that transport properties of oil are affected by wettability condition [11, 12]. For this reason, different wettability conditions are applied for analyzing the performance of CO_2-LPG EOR.

It has been identified that CO_2-LPG flood is an effective method for MMP reduction causing oil recovery enhancement through many experimental studies. Compositional model for CO_2-LPG EOR is necessary to investigate how gas transport affects MMP reduction and oil recovery enhancement. In this research, compositional fluid and multiphase simulation models are developed and injected LPG mole fraction and composition are optimized based on recent oil, CO_2, propane, and butane prices for maximum net present value (NPV).

NUMERICAL SIMULATION

Fluid Modeling

Fluid data of Weyburn reservoir is referred for NPV based solvent injection simulation. Weyburn reservoir, located in southeast Saskatchewan and operated by PanCanadian Petroleum Ltd., has reached its economic limit of production by waterflooding. The reservoir is a target for CO_2 miscible flooding to enhance oil recovery. The oil composition is shown in Table 1 and comparison between computed fluid model properties and actual fluid data of Weyburn reservoir is given in Table 2 [13]. Oil gravity, formation volume factor, and gas-oil ratio are calculated through regression process to match separator experimental data. Saturation pressure is also computed by regression process. Details of the calculation techniques for saturation pressure can be found in [14]. Acceptable match of computed properties from fluid model and Weyburn's data increases reliability of the fluid model for compositional simulation.

Table 1: Modeled fluid composition for Weyburn oil

Components	Mole fraction
N_2	0.0207
CO_2	0.0074
H_2S	0.0012
CH_4	0.0749
C_2H_6	0.0422
C_3H_8	0.0785
i-C_4 to n-C_4	0.0655
i-C_5 to n-C_5	0.0459
C_{6+}	0.6637
Total	1

Table 2: Comparison between properties of fluid model and Weyburn data

Parameters	Fluid model	Weyburn
Saturation pressure (psi)	688	713
Oil gravity (°API)	47	31
Formation volume factor (bbl/STB)	1.11	1.12
Gas-oil ratio (SCF/STB)	166	32
Minimum miscibility pressure (psi)	1,996	2,059

Phase behavior of fluid model was determined by Peng-Robinson EOS [15] with the reservoir oil and injected fluid composition. The PR EOS is given by

$$p = \frac{RT}{v-b} - \frac{a}{v(v+b)+b(v-b)},$$

(1)

Or in terms of Z factor,

$$Z^3 - (1-B)Z^2 + (A - 3B^2 - 2B)Z - (AB - B^2 - B^3) = 0,$$ (2)

and. $Z_c = 0.3074$

The EOS constants for pure components are given by

$$A = a\frac{p}{(RT)^2},$$

$$B = b\frac{p}{RT},$$

$$a = \Omega_a^o\frac{R^2T_c^2}{p_c}\alpha,$$

$$b = \Omega_b^o\frac{RT_c}{p_c},$$

$$\alpha = \left[1 + m\left(1 - \sqrt{T_r}\right)\right]^2, \tag{3}$$

Where

$$\Omega_{al}^o = 0.45724, \; \Omega_b^o = 0.07780,$$

And

$$m = 0.37464 + 1.54226\omega - 0.26992\omega^2. \tag{4}$$

Robinson and Peng [16] proposed a modified m for heavier components (w>0.49) as follows:

$$m = 0.3796 + 1.485\omega - 0.1644\omega^2 + 0.01667\omega^3. \tag{5}$$

Fugacity expressions are given by

$$\ln\phi_i = Z - 1 - \ln(Z - B)$$

$$-\frac{A}{2\sqrt{2}B}\left(\frac{B_i}{B} - \frac{2}{A}\sum_{j=1}^{N}y_iA_{ij}\right)\ln\left[\frac{Z + \left(1 + \sqrt{2}\right)B}{Z - \left(1 - \sqrt{2}\right)B}\right], \tag{6}$$

Where mixing rules are used for multicomponent fugacity expression as follows:

$$A = \sum_{i=1}^{N} \sum_{j=1}^{N} y_i y_j A_{ij},$$

$$B = \sum_{i=1}^{N} y_i B_i,$$

$$A_{ij} = \left(1 - k_{ij}\right) \sqrt{A_i A_j},$$

Where K_{ij} is binary-interaction parameters

Multiple mixing cell method [17] was applied to fluid model to estimate MMP between injected CO_2 and reservoir oil. Multiple mixing cell method follows the order below.

- Specify the reservoir temperature and an initial pressure.
- Calculate the tie-line length for each pressure step by using the equation below:

$$TL = \sqrt{\sum_{i=1}^{N_c} (x_i - y_i)^2},$$

(8)

Where N_c is the number of components and x_i and y_i are liquid and gas equilibrium compositions, respectively.

- Draw a tie-line length graph as a function of pressures.
- Perform a multiple-parameter regression of the minimum tie-line lengths to determine the exponent n in $TL^n = aP + b$ (power-law extrapolation). These parameters are determined when correlation coefficient exceeds 0.999.
- Determine the MMP when the power-law extrapolation gives zero of minimum tie-line length.

After generating the fluid model which has approximate MMP to Weyburn fluid, MMPs were computed between oil and LPGs. The composition of LPG is propane 63% and butane 37%, and the calculated MMPs are indicated in Table 3. MMPs of LPG (composition:

propane 100% and butane 0%) mole fraction 20% and 25% are 1,747 psi and 1,614 psi.

Table 3: MMP estimates for injection gas according to LPG mole fraction

LPG mole fraction (%)	MMP (psi)
0	1,996
5	1,995
10	1,825
15	1,820
20	1,412
25	1,354
30	1,046

Interfacial Tension Calculations

The equation for calculating interfacial tension in multicomponent systems is as follows [18]:

$$\sigma^{1/4} = \sum_{i=1}^{n_c} P_{ar_i} \left(x_i \rho_L - y_i \rho_g \right),$$

(9)

Where σ is the interfacial tension between liquid and gas phases (dyne/cm) and ρ_L and ρ_g are molar densities of liquid and gas phases (mole/cm³), respectively. The parachor (p_{ar_i}) is defined as follows:

$$P_{ar_i} = \xi CN_i,$$

(10)

Where

$$\xi = \begin{cases} 40, & CN \le 12, \\ 40.3, & CN > 12, \end{cases}$$

(11)

and CN is the carbon number of the components i.

Reservoir Modeling

The reservoir model was assumed as 2D model which is discretized into 33 × 33 × 1 grid blocks. Each grid block has dimension as 10 ft × 10 ft × 20 ft as shown in Figure 2. The model size is general one injector-one producer scale of 10-acre five-spot model [19]. This simulation study utilized homogeneous 2D areal model not considering heterogeneity and gas overriding effect. Without these effects, oil recovery can be governed only by displacement efficiency from LPG addition and can be expected near 100% [20].

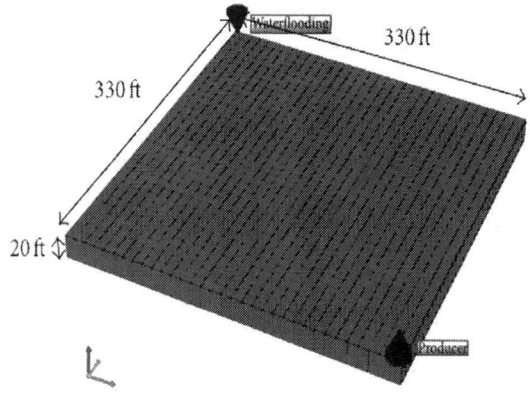

Figure 2: 3D view of simulation model.

Contact angle which is a determinant for wettability is defined by Young's equation as follows:

$$\cos\theta = \frac{\sigma_{os} - \sigma_{ws}}{\sigma_{ow}},$$

(12)

where $\sigma_{os,}$, $\sigma_{ws,}$ and $\sigma_{ow,}$ are oil-solid, water-solid, and oil-water interfacial tensions. As indicated in the above equation, if $\sigma_{os,}$ is greater than $\sigma_{ws,}$, θ is smaller than 90°, so the reservoir rock exhibits water-wet solid. The inverse case is oil-wet condition. Water-wet and oil-wet reservoirs have the constant porosity and isotropic permeability is also assumed. Reservoir initial conditions are shown in Table 4. The porosity, permeability, and relative permeability were gained from the same reservoir, and two different relative permeability curves (Figure 3) are used in this simulation for establishing different residual oil saturation and mobility [21]. The relative permeability curves are predicted by simulations. Simulation methods to predict relative permeability are already verified by previous studies [22, 23] and similar water relative curve can be found. Residual oil saturations of oil- and water-wet reservoirs are 18% and 15%, respectively.

Table 4: Properties of reservoir rock and fluids

Properties	Values
Depth (ft)	4,000
Pressure (psi)	2,000
Temperature (°F)	145
Permeability (md)	122
Porosity (%)	24
Oil saturation (s_o)	0.64
Water saturation (s_w)	0.36

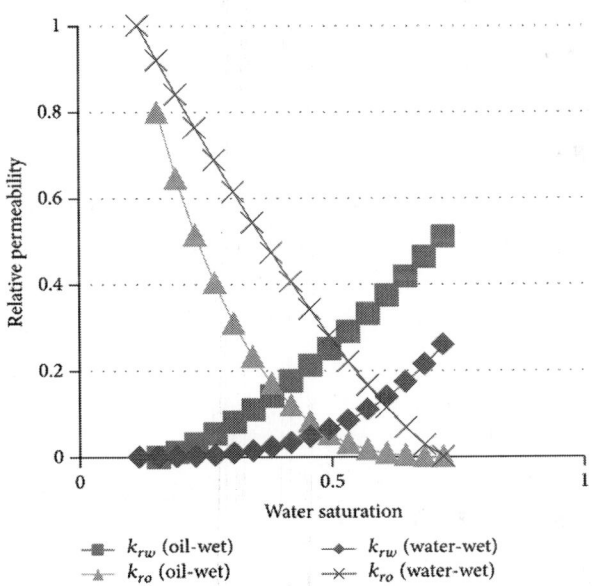

Figure 3: Relative permeability curves for different wettability conditions.

After waterflooding for three years, water-alternating CO_2 EOR and CO_2-LPG EOR were applied to water- and oil-wet reservoirs for ten years. WAG cycle of CO_2 and CO_2-LPG EOR is 1:1, and one cycle period is 6 months. Production pressure is 1,500 psi which is within a limitation of miscible condition by first or multiple contact miscibility process when added LPG concentration is larger than 20% (Figure 4). Injected LPG mole fraction and composition are indicated in Table 5.

Table 5: Operating conditions and injection design parameters

Properties	Values
Producing pressure at bottom hole (psi)	1,500
Total injection (PV)	1.5
Period (years)	10
WAG ratio	1:1
Injected LPG mole fraction (%)	0, 10, 15, 20, 25, and 30
Injected LPG composition (propane : butane)	100:0, 63:37, 37:63, and 0:100

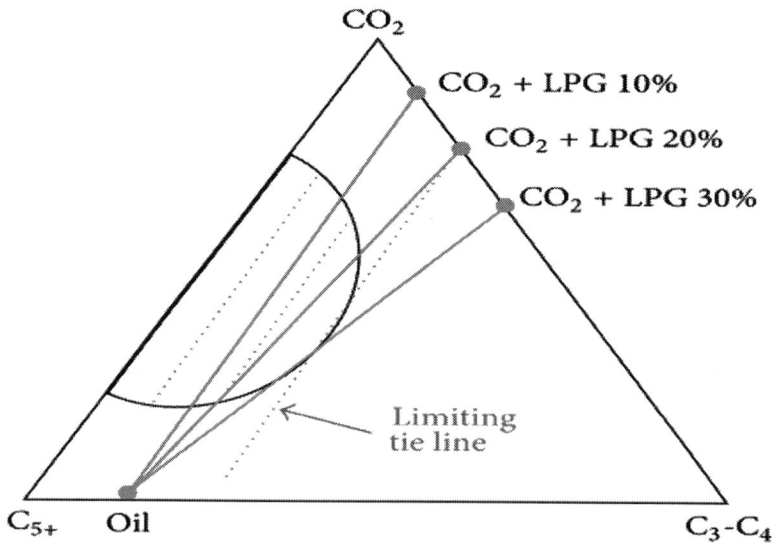

Figure 4: Ternary diagram for CO_2/LPG/Oil system at reservoir pressure 1,500 psi.

Net Present Value

The NPV of a time series of cash flows is defined as the sum of the present values. NPV considering prices of oil, CO_2, propane, and butane and costs of water injection and produced water handling is calculated by the following equation [24]:

$$NPV = \sum_{t=1}^{T} \frac{R_t}{(1+I)^t},$$

(13)

Where T is total production period (day), t is time, R_t is net profit at time t, and I is daily discount rate. I is estimated by yearly discount rate as

$$I = e^{\ln(1+\text{Yearly discount rate})/365} - 1,$$

(14)

where yearly discount rate is 10% and R_t is defined by the difference between the profit from oil production and total investment costs at time :

$$R_t = Q_o P_o - \left(Q_{CO_2} P_{CO_2} + Q_{C_3} P_{C_3} \right.$$
$$\left. + Q_{C_4} P_{C_4} + Q_{w_1} P_{w_1} + Q_{w_2} P_{w_2} \right),$$

(15)

where Q_o, Q_{CO_2}, Q_{C_3}, Q_{C_4}, Q_{w_1}, and Q_{w_2} are oil production rate (bbl/day), CO_2 injection rate (lb/day), propane injection rate (lb/day), butane injection rate (lb/day), water injection rate (bbl/day), and water production rate (bbl/day). Parameters P_o, P_{CO_3}, P_{C_3}, P_{C_4}, P_{w_1}, and P_{w_2} are oil price ($/bbl), CO_2 price ($/lb), propane price ($/lb), butane price ($/lb), water injection cost ($/bbl), and produced water handling cost ($/bbl). All values of parameters for NPV calculation are shown in Table 6 [25, 26].

Table 6: Economic parameters for optimal design

Parameters	Values
Oil ($/bbl)	80
CO_2 ($/ton)	80
Propane ($/ton)	800
Butane ($/ton)	850
Water injection ($/bbl)	0.25
Produced water handling ($/bbl)	1.5

RESULTS AND DISCUSSION

Oil Production

The aim of this study is to confirm the effectiveness of water-alternating CO_2-LPG EOR process in oil recovery for different reservoirs. The performance of CO_2-LPG injection process has been compared with that of CO_2WAG process. LPG is composed of 63% propane and 37% butane. Results of oil recovery with various LPG concentrations are indicated in Figure 5. Increased oil recoveries for oil- and water-wet reservoirs by CO_2-LPG flood are 46% and 22%. For both wettability conditions, the higher LPG mole fraction is injected, the more oil is produced. However, significant differences are not found if LPG mole fraction is greater than 25%. The tendency is also identified by experimental results in the literature [7]. To detect the influence of LPG composition in the injected fluid, oil recoveries with different ratio of propane and butane (LPG 15%) are shown in Figure 6. Figure 6 shows that the higher fraction of butane causes more enhanced oil recovery. Increments of oil recovery for oil- and water-wet reservoirs are 25% and 15% as compared with CO_2 EOR. This phenomenon was already revealed by the experimental study and it was explained that the result is because of higher mole weight [27].

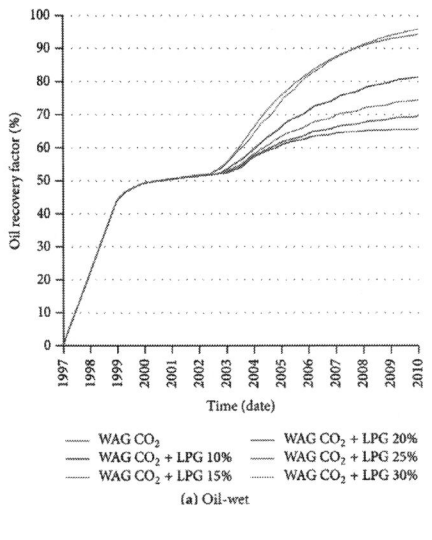

(a)

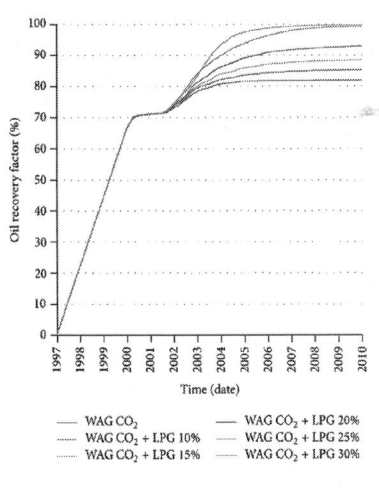

(b)

Figure 5: Oil recovery factors with LPG mole fraction of injection gas for different wettability conditions (composition of LPG: propane 63%, butane 37%).

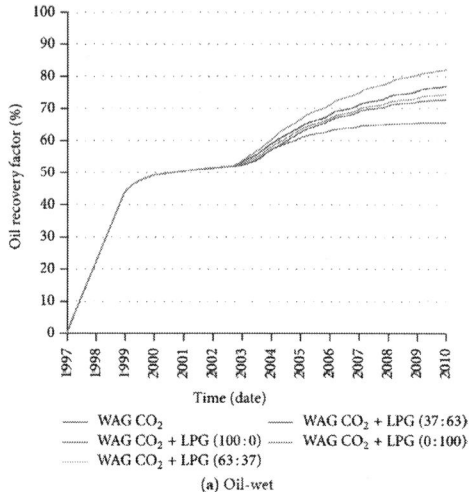

(a)

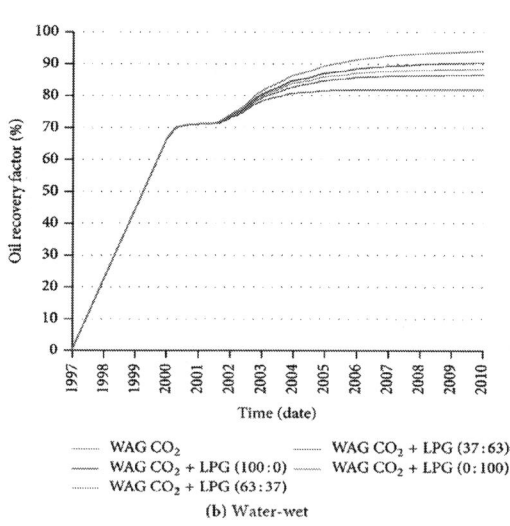

(b)

Figure 6: Oil recovery factors with LPG composition for different wettability conditions (LPG mole fraction of injection gas: 15%).

When reservoir oil and injected gas are miscible, gas saturation decreases further than immiscible condition (Figure 7). Injected gas reached production well at around 2004, so the gas saturation of WAG CO_2 case increased abruptly. However, in case of WAG CO_2 + LPG 30%, the gas saturation did not increase even though injected gas already reached production well. It indicates that miscible condition reduces gas saturation. The reduction of gas saturation causes a decrease in gas relative permeability (Figure 8). As both gas saturation and relative permeability decline, liquid saturation and relative permeability increase, which leads to the enhancement of oil recovery.

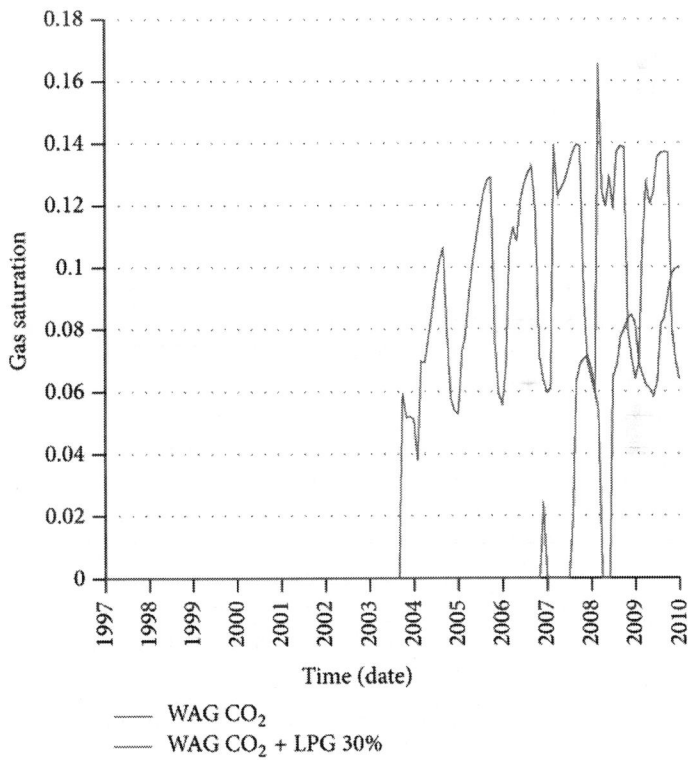

WAG CO_2

WAG CO_2 + LPG 30%

Figure 7: Gas saturation with LPG mole fraction of injection gas near production well for oil-wet condition.

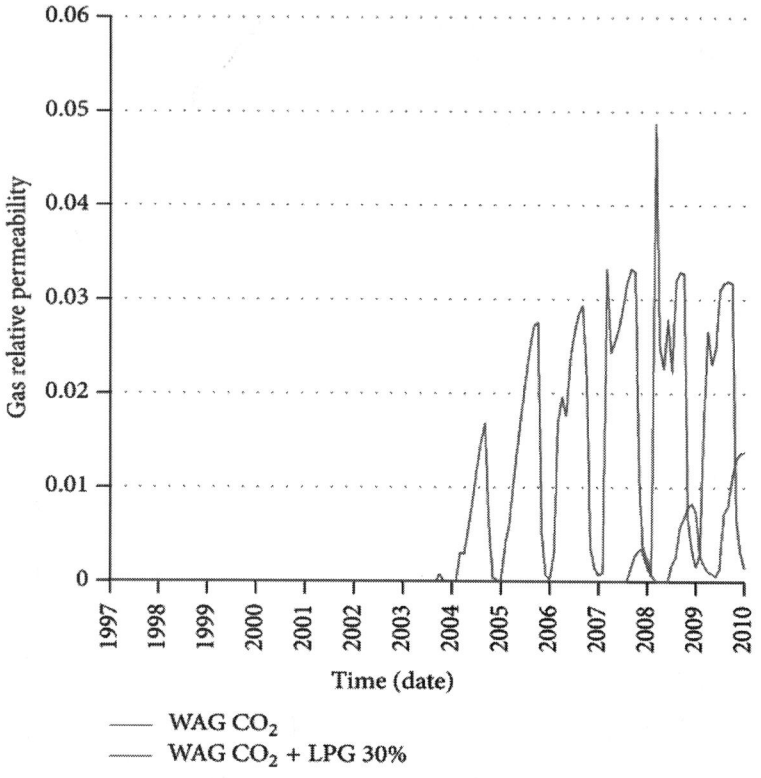

Figure 8: Gas relative permeability with LPG mole fraction of injection gas near production well for oil-wet condition.

The addition of LPG to CO_2 stream is more effective to lower interfacial tension between oil and gas phases. In particular, if the reservoir is in miscibility condition, interfacial tension reaches zero [1]. As shown in Figure9 (a), the swept zone is left in nonzero interfacial area, so reservoir is not in miscible condition by only CO_2 injection. In contrast, the addition of LPG to CO_2 stream as 20% brings the swept area into zero interfacial zone indicating miscible condition (Figure 9(b)). Zero interfacial tension indicates that oil and gas become single-phase, so it flows easier than two-phase fluid.

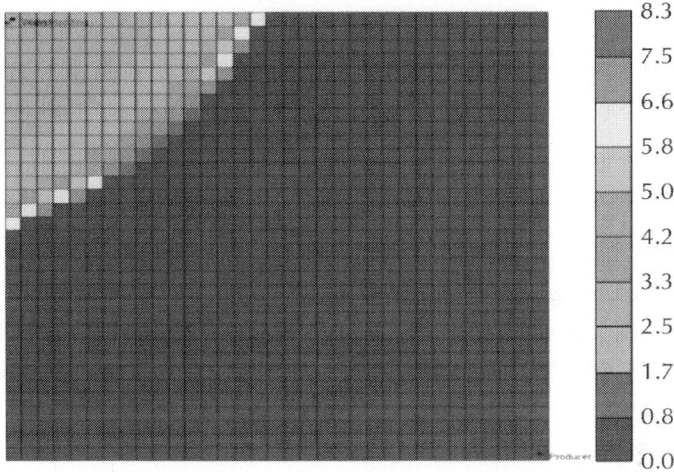

(a)

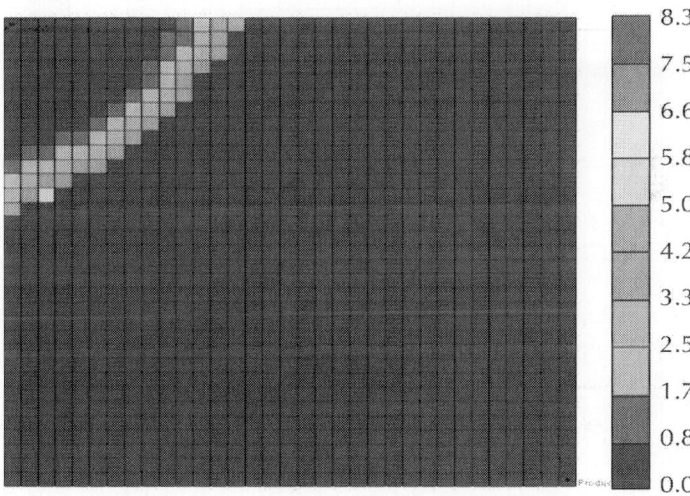

(b)

Figure 9: Interfacial tension between oil and gas phases with injected gas mole fraction after six months of gas injection (LPG composition: 63% propane and 37% butane).

If more butane content is injected than propane, more oil recovery is expected because of its higher molecular weight. It was proved that butane is much more effective in MMP reduction [27]. The addition of alkane solvents to the CO_2 stream accelerates the process of reducing oil viscosity; thus, it leads to higher oil recovery [4]. To compare the aspect of oil recovery by injected LPG composition, oil saturation in reservoir is shown in Figure 10. Figure 10 indicates the oil saturation when LPG mole fraction is 25% for oil-wet reservoir after 6 months from the end of waterflooding. In case of 100% propane, oil saturation near injection well is zero because of immaculate expulsion and it is 0.5 at the fore-end of injected fluid (Figure 10(a)). In case of 100% butane, zero zone of oil saturation is more widespread with near wellbore as a center. Furthermore, oil saturation at the fore-end is 0.7 which is higher than that in the case of 100% propane because more oil is displaced from wider area (Figure 10(b)).

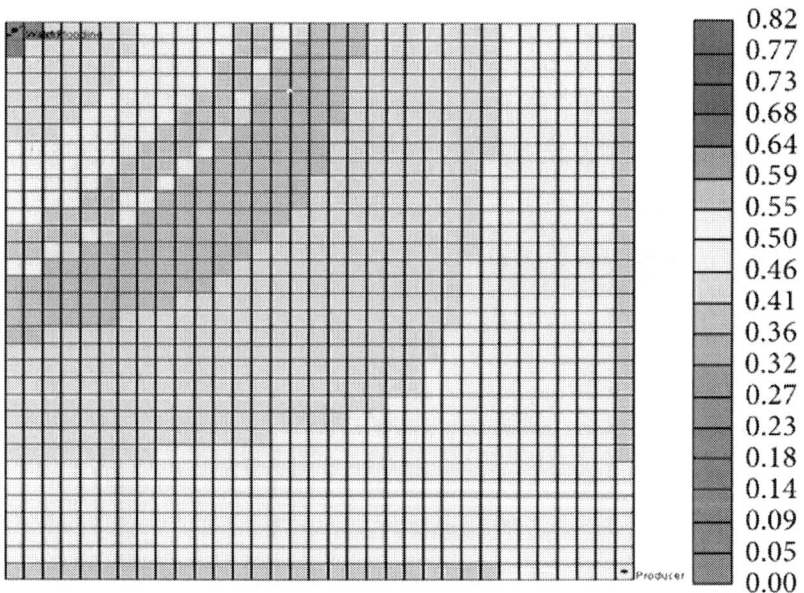

(a)

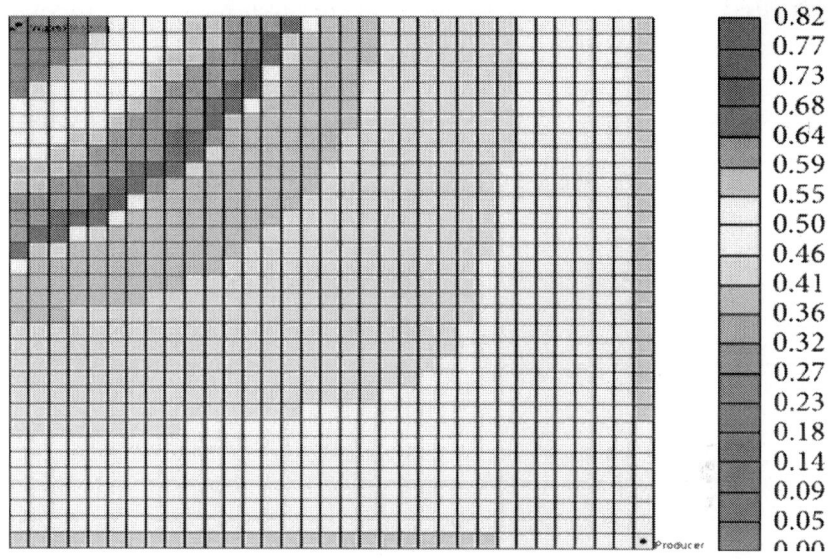

	0.82
	0.77
	0.73
	0.68
	0.64
	0.59
	0.55
	0.50
	0.46
	0.41
	0.36
	0.32
	0.27
	0.23
	0.18
	0.14
	0.09
	0.05
	0.00

(b)

Figure 10: Oil saturation with LPG composition after six months of gas injection (LPG volume fraction of injection gas: 15%).

Tables 7 and 8 indicate the amount of increase in maximum NPV after water flooding for different wettability conditions. NPVs are calculated according to LPG concentration and composition. The maximum NPV increments by CO_2 WAG are 12% and 13% for oil- and water-wet reservoirs. As shown in Tables 7 and 8, the maximum value is 24.1% (LPG 25%: propane 63%, butane 37%) for oil-wet reservoir and 17.0% (LPG 20%: propane 0%, butane 100%) for water-wet reservoir. When LPG mole fraction is less than 15% and 20% (propane 100%), maximum increase in NPV is less than CO_2 WAG. These cases are in immiscible condition, so oil recovery is not high compared to the economic feasibility of LPG. Injected fluid and reservoir oil are in miscible condition, and maximum NPV improvements are higher than those of CO_2 WAG cases. Maximum NPV increment by CO_2-LPG EOR occurred for two different wettability conditions, but the effect in oil-wet reservoir is better than in water-wet reservoir because of higher residual oil saturation after waterflooding

Table 7: Maximum NPV improvements with LPG mole fraction and composition for oil-wet reservoir (base case: $2,652,887)

	Maximum NPV improvements (%)				
	LPG 10%	**LPG 15%**	**LPG 20%**	**LPG 25%**	**LPG 30%**
Propane 100% Butane 0%	10.7	11.0	11.8	12.8	20.0
Propane 63 Butane 37%	9.3	10.3	13.0	24.1	19.2
Propane 37% Butane 63%	9.0	10.9	18.4	19.2	15.4
Propane 0% Butane 100%	9.2	13.8	22.5	17.3	13.0

Table 8: Maximum NPV improvements with LPG mole fraction and composition for water-wet reservoir (base case: $3,387,572)

	Maximum NPV improvements (%)				
	LPG 10%	**LPG 15%**	**LPG 20%**	**LPG 25%**	**LPG 30%**
Propane 100% Butane 0%	12.2	12.0	12.3	12.6	15.5
Propane 63% Butane 37%	11.8	12.2	13.1	16.0	16.7
Propane 37% Butane 63%	11.8	12.5	15.6	16.7	14.2
Propane 0% Butane 100%	11.7	13.6	17.0	15.4	12.4

CONCLUSIONS

In this study, water-alternating CO_2-LPG EOR simulation model was developed. To examine the efficiency of CO_2-LPG EOR considering oil, CO_2, and LPG prices, extensive simulations have been performed for different wettability conditions and the following conclusions have been drawn.

- When LPG concentration is 30% and composition of butane is 100%, oil recovery increased by 46% and 25% for oil-wet reservoir. When LPG concentration is 30% and butane composition is 100%, the maximum increasing amounts are 22% and 15% in case of water-wet reservoir. As injected LPG concentration and butane composition increased, significantly enhanced oil recovery was observed from the reduction of MMP and interfacial tension. Oil recovery for different wettability by CO_2-LPG EOR has become close to 100%.

- When LPG concentration is 25% and butane composition is 37%, maximum NPV improvement is 24.1% for oil-wet reservoir. When LPG concentration is 20% and butane composition is 100%, maximum NPV improvement is 17.0% for water-wet reservoir. For both oil- and water-wet reservoirs, when LPG concentrations are 10%, 15%, 20% (propane 100%), and 25% (propane 100%), the reservoir condition is immiscible and maximum NPV increment is lower than CO_2 WAG process. When LPG concentration is higher than 20% (miscible condition), maximum NPV improved and optimum LPG concentration and composition exist for maximum NPV improvement.

- CO_2-LPG EOR can be applicable in low pressure reservoirs that CO_2 is not miscible. LPG addition to CO_2 stream can appreciably improve oil recovery by zero interfacial tension bringing miscible condition. Moreover, the optimization of LPG concentration and composition is absolutely necessary for economic feasibility. The necessity of optimization is required more in oil-wet reservoir due to better performance of displacement efficiency.

ACKNOWLEDGMENTS

This work was supported by the Energy Efficiency & Resources Core Technology Program of the Korea Institute of Energy Technology Evaluation and Planning (KETEP) granted financial resource from the Ministry of Trade, Industry & Energy, Republic of Korea (no. 20122010200060).

REFERENCES

1.   D. Makimura, M. Kunieda, Y. Liang, T. Matsuoka, S. Takahashi, and H. Okabe, "Application of molecular simulations to CO_2-enhanced oil recovery: phase equilibria and interfacial phenomena,"Society of Petroleum Engineers, vol. 18, no. 2, pp. 319–330, 2013. ·

2.   F. I. Stalkup Jr., Miscible Displacement, Society of Petroleum Engineers of AIME, Dallas, Tex, USA, 2nd edition, 1983.

3.   K. Talbi, T. M. V. Kaiser, and B. B. Maini, "Experimental investigation of CO_2-based VAPEX for recovery of heavy oils and bitumen," Journal of Canadian Petroleum Technology, vol. 47, no. 4, pp. 29–36, 2008.

4.   H. Li, S. Zheng, and D. Yang, "Enhanced swelling effect and viscosity reduction of solvent(s)/CO_2/heavy-oil systems," SPE Journal, vol. 18, no. 4, pp. 695–707, 2013.

5.   L. W. Lake, Enhanced Oil Recovery, Society of Petroleum Engineers, Richardson, Tex, USA, 2010.

6.   R. B. Alston, G. P. Kokolis, and C. F. James, "CO_2 minimum miscibility pressure: a correlation for impure CO_2 streams and live oil systems," Society of Petroleum Engineers Journal, vol. 25, no. 2, pp. 268–274, 1985. · ·

7.   N. Kumar and W. D. V. Gonten, "An investigation of oil recovery by injecting CO_2 and LPG mixtures," in Proceedings of the 48th Annual Fall Meeting of the Society of Petroleum Engineers of AIME, SPE 4581, Las Vegas, Nev, USA, September 1973.

8.   P. Y. Zhang, S. Huang, S. Sayegh, and X. L. Zhou, "Effect of CO_2 impurities on gas-injection EOR processes," in Proceedings of the SPE/DOE Symposium on Improved Oil Recovery, SPE 89477, Tulsa, Okla, USA, April 2004.

9.   E. M. E. Shokir, "CO_2-oil minimum miscibility pressure model for impure and pure CO_2 streams," inProceedings of the Offshore Mediterranean Conference and Exhibition (OMC '07), Ravenna, Italy, March 2007.

10.   T. W. Teklu, N. Alharthy, H. Kazemi, X. Yin, and R. M. Graves, "Hydrocarbon and non-hydrocarbon gas miscibility with light

oil in shale reservoirs," in Proceedings of the SPE Improved Oil Recovery Symposium, SPE 169123, Tulsa, Okla, USA, April 2014.

11. J. Cai, X. Hu, D. C. Standnes, and L. You, "An analytical model for spontaneous imbibition in fractal porous media including gravity," Colloids and Surfaces A: Physicochemical and Engineering Aspects, vol. 414, pp. 228–233, 2012. · ·

12. J. Cai, E. Perfect, C.-L. Cheng, and X. Hu, "Generalized modeling of spontaneous imbibition based on hagen-poiseuille flow in tortuous capillaries with variably shaped apertures," Langmuir, vol. 30, no. 18, pp. 5142–5151, 2014. · ·

13. R. K. Srivastava, S. S. Huang, and M. Dong, "Laboratory investigation of Weyburn CO_2 miscible flooding," Journal of Canadian Petroleum Technology, vol. 39, no. 2, pp. 41–51, 2000.

14. L. X. Nghiem, Y.-K. Li, and R. A. Heidemann, "Application of the tangent plane criterion to saturation pressure and temperature computations," Fluid Phase Equilibria, vol. 21, no. 1-2, pp. 39–60, 1985. · ·

15. D.-Y. Peng and D. B. Robinson, "A new two-constant equation of state," Industrial and Engineering Chemistry Fundamentals, vol. 15, no. 1, pp. 59–64, 1976. · ·

16. D. B. Robinson and D. Y. Peng, "The characterization of the heptanes and heavier fractions," Research Report 28, Gas Processors Association, Tulsa, Okla, USA, 1978.

17. K. Ahmadi and R. T. Johns, "Multiple-mixing-cell method for MMP calculations," SPE Journal, vol. 16, no. 4, pp. 733–742, 2011. · ·

18. R. C. Reid, J. M. Prausnitz, and T. K. Sherwood, The Properties of Gases and Liquids, McGraw-Hill, New York, NY, USA, 1977.

19. G. P. Willhite, Waterflooding, Society of Petroleum Engineers, Richardson, Tex, USA, 1986.

20. G. F. Teletzke, P. D. Patel, and A. L. Chen, "Methodology for miscible gas injection EOR screening," inProceedings of the SPE International Improved Oil Recovery Conference in Asia Pacific (IIORC ‹05), SPE 97650, pp. 315–325, Kuala Lumpur, Malaysia, December 2005.

21. M. Delshad, N. F. Najafabadi, G. A. Anderson, G. A. Pope, and K. Sepehrnoori, "Modeling wettability alteration in naturally

fractured reservoirs," in Proceedings of the 15th SPE/DOE Improved Oil Recovery Symposium, vol. 2 of SPE 100081, Tulsa, Okla, USA, April 2006.

22. B. Xiao, J. Fan, and F. Ding, "Prediction of relative permeability of unsaturated porous media based on fractal theory and Monte Carlo simulation," Energy and Fuels, vol. 26, no. 11, pp. 6971–6978, 2012. · ·

23. B. Xiao, J. Fan, and F. Ding, "A fractal analytical model for the permeabilities of fibrous gas diffusion layer in proton exchange membrane fuel cells," Electrochimica Acta, vol. 134, pp. 222–231, 2014. · ·

24. S. Salem and T. Moawad, "Economic study of miscible CO_2 flooding in a mature waterflooded oil reservoir," in Proceedings of the SPE Saudi Arabia Section Annual Technical Symposium and Exhibition, SPE 168064, Al-Khobar, Saudi Arabia, May 2013.

25. C. L. Liao, X. W. Liao, X. L. Zhao et al., "Study on enhanced oil recovery technology in low permeability heterogeneous reservoir by water-alternate-gas of CO_2 flooding," in Proceedings of the SPE Asia Pacific Oil and Gas Conference and Exhibition, SPE 165907, Jakarta, Indonesia, October 2013.

26. OPIS Europe LPG Report, 2014, http://www.opisnet.com.

27. R. S. Metcalfe, "effects of impurities on minimum miscibility pressures and minimum enrichment levels for CO_2 and rich-gas displa cements," SPE Journal, vol. 22, no. 2, pp. 219–225, 1982. · ·

3

Petroleum Reservoir Uncertainty Mitigation through the Integration with Production History Matching

Gustavo Gabriel Becerra[I], Célio Maschio[II], and Denis José Schiozer[III]

[I]Petrobras Energia S.A., Cenpes, 21941-915 Rio de Janeiro, RJ, Brazil

[II]Universidade Estadual de Campinas – UNICAMP, FEM/DEP/UNISIM – CEPETRO, 13083-970 Campinas, SP, Brazil

[III]Universidade Estadual de Campinas – UNICAMP, FEM/DEP/UNISIM – CEPETRO, 13083-970 Campinas, SP, Brazil

ABSTRACT

This paper presents a new methodology to deal with uncertainty mitigation using observed data, integrating the uncertainty analysis and the history matching processes. The proposed method is robust and easy to use, offering an alternative way to traditional history matching methodologies. The main characteristic of the methodology is the use of observed data as constraints to reduce the uncertainty of the reservoir parameters. The integration of uncertainty analysis with history matching naturally yields prediction under uncertainty. The workflow permits to establish a target range of uncertainty that characterize a confidence interval of the probabilistic distribution curves around the observed data. A complete workflow of the proposed methodology was carried out in a realistic model based on outcrop data and the impact of the uncertainty reduction in the production forecasting was evaluated. It was demonstrated that for complex cases, with a high number of uncertain attributes and several objective-function, the methodology can be applied in steps, beginning with a field analysis followed by regional and local (well level) analyses. The main contribution of this work is to provide an interesting way to quantify and to reduce uncertainties with the objective to generate reliable scenario-based models for consistent production prediction.

INTRODUCTION

The geological, reservoir, economic and technologic uncertainties influence the management decisions of hydrocarbon reserves and of future development plans. Consequently, the quantification of the impact of these uncertainties provides an increased reliability of this process.

The uncertainty term states the degree of knowledge about the properties of the system under analysis. The risk concept indicates the objective-functions (OF) variability of the problem, obtained from the probability analysis of the possible scenario-based models. In the context of this work, the OF indicates the misfit between the observed production and pressure data and the simulated data of the corresponding models. The cumulative distribution of the objective-function probabilities is a density curve, known as uncertainty curve, which allows determining the history matching quality for the analyzed possible models.

The scarcity of quality information makes the construction of a dynamic model difficult, making it necessary its calibration derived from the productive response measured in the field. The history matching is an inverse problem, in which different combinations of the reservoir's parameter values can lead to acceptable responses, especially when the degree of uncertainty of these parameters is high. The problem tends to worsen in the cases when the history period is short. Even though different solutions provide reasonable confidence comparing with observations, any one of them could produce a different prediction, leading to a range of distinct responses.

The methodology used in this paper leads to the detection of calibrated models within the range of defined acceptability. The integration is made gradually, proceeding through stages (global, regional and local), for the different attributes and the identified objective-functions (OF). The objective is seeking to reduce the occurrence probabilities of those scenarios that do not present a good matching and, consequently, increase the probabilities of the models which have performed close to the history. This paper presents methods that make possible a redefinition of the values of the studied uncertain attributes, allowing a reduction of the uncertainty in the history matching stage as well as in the prediction period.

The uncertainty inherent to dynamic modeling of a reservoir depends on several factors. One of them is a consequence of the model's own error in trying to represent a reality. Other factors are caused by random nature and insufficient static and dynamic data. The uncertainties are analyzed taking into account that knowledge of the reservoir is only partial, using, in the initial phases of exploration and discovery of a field, indirect information, having few, sparse direct data. From the field development up to its abandonment, new information about the reservoir is added, but the knowledge is always partial and incomplete. Thus, it is necessary to incorporate a probabilistic approach in the history matching and predictions of production with uncertainty.

Traditionally, the uncertainty analysis is applied in the initial stages or in the prediction phase; however, the advance was small in the use of this analysis in history matching studies. Obtaining the best deterministic matching is not the target of the proposed methodology, but rather reflecting on how the history data makes possible the mitigation of uncertainties.

The objective of this paper is to apply and improve, in more complex reservoir model, the methodology proposed by Maschio et al. (2005) and Moura Filho (2006), originally developed in a simple model. The static and dynamic data, detected in the uncertainties analysis workflow, are included through a consistent methodology that permits integration of probabilistic analyses of the uncertain attributes with the history matching process.

LITERATURE

The first papers presented in the technical literature combining probability analysis procedures of static and dynamic data with a variety of scenarios date from the 1990's. The multi-disciplinary approach to history matching combined with uncertainty analysis is rather recent (approximately 8 to 10 years) and there is a variety of treatments in the literature. Roggero (1997); Christie et al. (2002) and Kashib and Srinivasan (2006) proposed methods based on conditional probabilities, following the Bayesian formalism, to update the distribution of geologic attributes taking into consideration the additional information contained in the dynamic responses of the observed variables.

The combination of geostatistical modeling and the recorded history values is discussed by Bissel (1997); Bennett and Graf (2000) and Jenni et al. (2004). The uncertainties of fields in production are estimated by means of generating multiple reservoir models and evaluating the history matching through the respective gradient information, demanding a large computation effort. The practical use can be limited depending on the complexity of the models. Zabalza-Mezghani et al. (2004) present several options for the uncertainties management based on techniques of experimental design, construction of proxy-models and the combined use of geostatistics. The method consists in obtaining multiple history matching considered probabilistically equivalent by the stochastic proximity, and then extrapolated for the prediction under uncertainty analysis.

Lépine et al. (1999) propose a practical method, although restrictive, to calculate the effects of the uncertainties during the prediction period. From a single history matched simulation model, using gradient minimization techniques, the base values of attributes that permit the solution are slightly disturbed. Then, the modification of the selected gradients allows a range of possible future production profiles to be obtained. Landa and Guyaguler (2003) proposed the use of the gradient information of uncertainty attributes to determine the influence of the uncertainties and the subsequent construction of response surface at the end of the history period. Proxy-models are also used to reduce the computational effort required by the combination of a large quantity of uncertainty attributes to reach the representative models. Along the same line are the works of Manceau et al. (2001).

The joined matching of production data with seismic attributes is the line of study begun by Guérillot and Pianelo (2000). Litvak et al. (2005) presented an article for the estimation of the degree of prediction variation by means of production and seismic data. The neighborhood algorithm was applied to select the matching parameters in each simulation. Varela et al. (2006) used the seismic amplitude data and analyzed its influence on production performance to reduce the prediction uncertainties. When the authors evaluated the range of production predictions, it was observed that the seismic amplitude data do not improve uniformly the variability of predictions for water breakthrough time in production wells.

The use of statistical methods is another analytical line. Gu and Oliver (2004) applied the Kalman filter method to obtain automatic multiple history matching for subsequent estimation of the predictions uncertainty. Alvarado et al. (2005) pointed out the importance of quantification of uncertainty in production predictions. A procedure that considers probability distribution of the prediction period based on the quality and weight attributed to the matching of a defined objective-function for the history period was proposed. Other papers along the same research topic are from Williams et al. (2004) and Ma et al. (2006). Queipo et al. (2002) present a methodology based on the use of artificial neural networks on efficient global optimization, for the calculation of the spatial distribution of permeability and porosity in heterogeneous reservoirs with multiple fluids through the calibration of available static and dynamic data. Reis (2006) also use artificial neural networks to combine risk analysis with history matching.

Based on the use of optimization algorithms, Nicotra et al. (2005) and Rotondi et al. (2006) showed methods of production prediction and uncertainty quantification using neighborhood algorithms, consisting of stochastic sampling algorithms, in search of an acceptable matching of the observed data. Also, using the neighborhood algorithm in conjunction with a geostatistical multiple-point process was the suggestion of Suzuki and Caers (2006), in whose paper each scenario was quantitatively described by a training image and a geological model execution, both stochastically generated.

From the bibliographic review, it can be deduced that the combined analysis of uncertainty and risk with history matching is a subject that has various, recent approaches. In the methodology showed in this paper, improved by Becerra (2007), the main differences in relation to the discussed methods are centered on the techniques of uncertainty quantification, on the OF used and on the way the degree of knowledge in certain areas of the reservoir is conditioned through the observed data.

METHODOLOGY

The main idea is to reduce the uncertainties as much as possible within boundaries set by the quantity and quality of the observed data. Consequently, the conditioned probabilistic analysis allows a quantitative integration approach. Three methods are presented, based on probability redistribution. In Method 1, there is a change in the initial probabilities assigned to the levels of uncertainty of the attributes. In Method 2, those uncertainty levels that produce great mismatch are discarded, reducing the number of possible scenarios. Method 3 implies the use of acceptance and evaluation criteria that conduct a reduction of the uncertain attributes variation range considered. The proposed methodology is more appropriate for petroleum fields in intermediate stages of production, in which a reasonable quantity of information is available, but, even so, a high degree of uncertainty exists in the description of the reservoir.

Original Methods

Several scenarios of the reservoir are obtained from the combinations of the most important uncertain attributes. Fig. 1(a) shows a general view of the procedure. The upper left frame illustrates an example of an uncertain attribute represented by a probability density function with three discrete levels. The graph also illustrates the probability redefinition of the discrete levels. The extreme values of the levels represent the initial variation range associated with a probability distribution. The lower left frame shows examples of the obtained cumulative probability curves, being that the central vertical line represents the history data. The frames to the right present the redefinition of the distributions and the effect on the production profile during the history and prediction period.

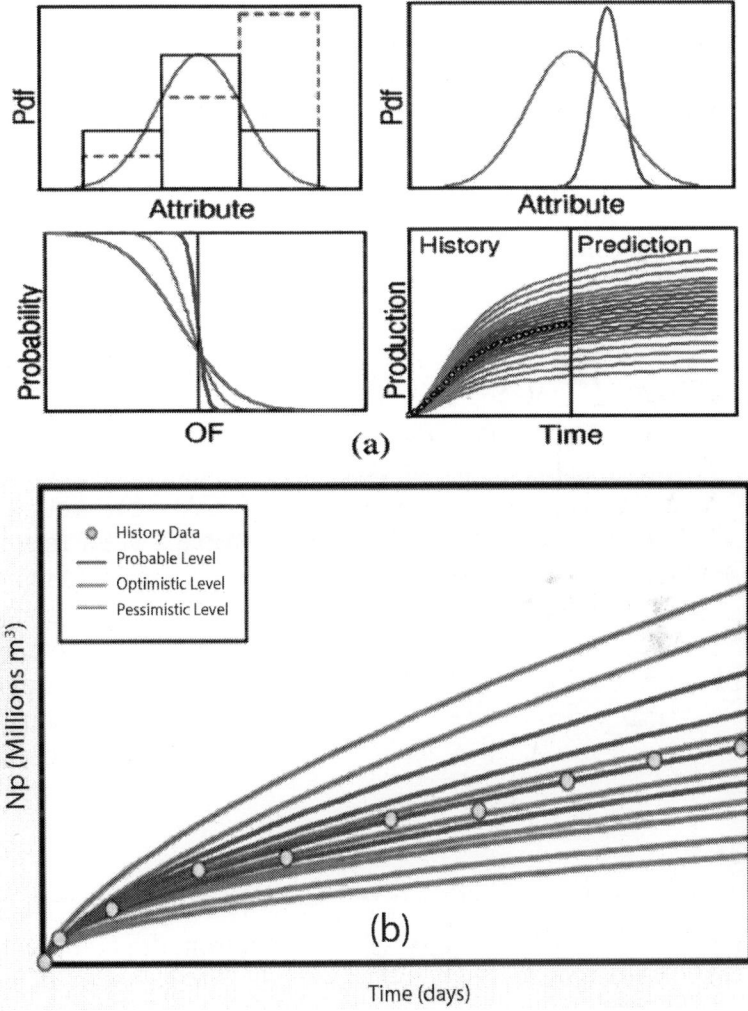

Figure 1: General aspect of the methodology (a) and examples production profiles (b).

The uncertainty quantification was carried out through derivative tree technique using reservoir simulation (Maschio et al., 2005); however, other techniques could be used (neural networks, experimental design combined with surface response, Monte Carlo simulation, etc). The levels of the uncertain attributes are combined, such that each branch of the tree results in a different simulation model. Thus, b^a models are generated, where 'b' is the number of levels and 'a' the number of attributes (Schiozer et al., 2005). For example, for four attributes each with 3 levels of uncertainty, the total number of simulations will be $3^4 = 81$. The inclusion of one more variable, also with three levels, elevates the number to $3^5 = 243$. This makes evident the importance of sensitivity analysis, in order to identify the more critical uncertain attributes and to limit the total number of simulations.

The OF is defined according to the following equations:

$$OF = \frac{D}{|D|} D_s$$

(1)

Where:

$$D = \sum_{i=1}^{N} (d_i^{obs} - d_i^{sim})$$

(2)

And

$$D_s = \sum_{i=1}^{N} (d_i^{obs} - d_i^{sim})^2$$

(3)

In the above equations, N is the number of observed data. The quotient D/|D| in Eq. (1) defines the sign of mismatching, indicating the position of the simulated data in relation to the observed data, an important concept for the next steps.

Method 1 uses the deviation distances calculated between the simulation models and the observed data for redistributing the probabilities of the attribute levels. The new probability for each level is calculated in accordance with the following equation:

$$P_n = \frac{\left(1/|S_n|\right)\left|D_n^{-1}\right|}{\sum_{L=1}^{k}\left(1/|S_n|\right)\left|D_n^{-1}\right|}$$

(4)

The subscript n identifies one of the discrete levels considered (0, 1 or 2 in the case of 3 levels, k = 3), while D_n and S_n are calculated by means of:

$$D_n = \left(\sum_{j=1}^{Mn} D_S\right)$$

(5)

And

$$S_n = \frac{D_n}{|D_n|}$$

(6)

In Eq. (4), Eq. (5) and Eq. (6), k is the number of discrete levels of the analyzed attribute, M_n is the number of models referred to the level n and the term D_n is the sum of deviation distances squared (D_s) of the M_n models during the history period considered. S_n factor represents a concept introduced as a measure of symmetry. It provides a greater probability value for those models better distributed around the production history curve. The sum (from $j = 1$ to M_n), in Eq. (5) is a global indicator of deviation above or below the values observed in the scenarios corresponding to the level n.

Consequently, the value of S_n varies between -1 and $+1$, zero being the value that indicates a curve distribution centered with respect to the history data. The value -1 indicates that all the curves are above the history data, and $+1$ that the curves are below the same. From the previous affirmation, it can be deduced that values close to zero have greater influence on the calculation of the respective value of P_n. In Eq. (4), the factor $(1/S_n)$ represents the degree of relative importance or weight of the group of curves for a given level. In the original work, a limitation for this factor is considered, with a maximum value of five, to avoid attributing very high weights and, consequently, to avoid excessive influence of the same:

$$1 \leq \frac{1}{|S_n|} \leq 5$$

(7)

From Eq. (7), the value of module $|S_n|$ varies in the interval from 0.2 to 1. Figure 1(b) exemplifies the distribution of the observed data with respect to the curves of possible reservoir models classified according to an uncertain attribute. The yellow points represent the production history. The curves in red are all located on the same side (below) of the history data, thus they present an S value equal to $+1$. The curves in green and in blue, however, are distributed around the production history and, for this reason, present S values that vary from -1 to $+1$. Even so, because the group of curves in blue presents greater symmetry around the history, their respective factor S value is closer to zero.

Figure 2 illustrates the aim of Method 1. The example schematizes the theoretic curves obtained from 9 scenarios derived from the combinations of two defined attributes with three defined levels. The three groups of curves represent the combination of the three levels of attribute A (A_0, A_1 and A_2) with each level of attribute B. Level A_2 receives the greatest probability because of the proximity of the corresponding models to the observed data. In the opposite direction, level A1 has a lower probability.

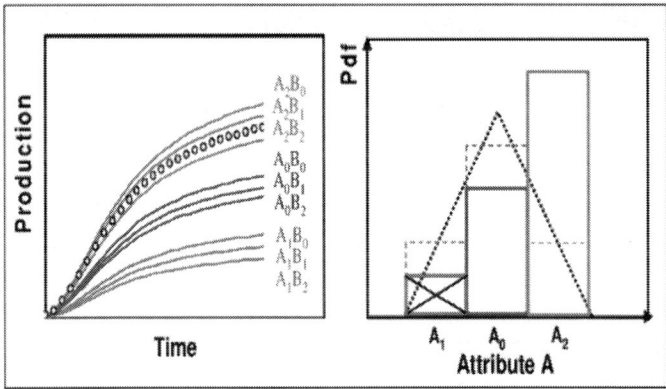

Figure 2: Schematic representation of Methods 1 and 2.

Method 2 consists of the elimination of one or more uncertainty levels of the attribute being considered and a redistribution of the probabilities resulting from this elimination. For a discrete level to be eliminated, it must satisfy the conditions expressed in Eq. (8) and Eq. (9).

$$\left| S_n \right| = 1$$

(8)

And

$$P_n \le 10\%$$

(9)

If the simulated curves are entirely asymmetric with respect to the history and if the occurrence probability of the level is less than 10%, this level is eliminated and the probabilities are redistributed to the remaining levels of that parameter. Considering the example of Fig. 2, level A1 is discarded and the values of the occurrence probabilities of the remaining levels are recalculated.

Finally, Method 3 consists in the redefinition of the uncertainty levels, in conformity with the curve distribution of the models relative to each attribute level. Following the example of Fig. 3(a), the new levels are calculated by Eq. (10) and Eq. (11).

$$Ls^N = Ls$$

(10)

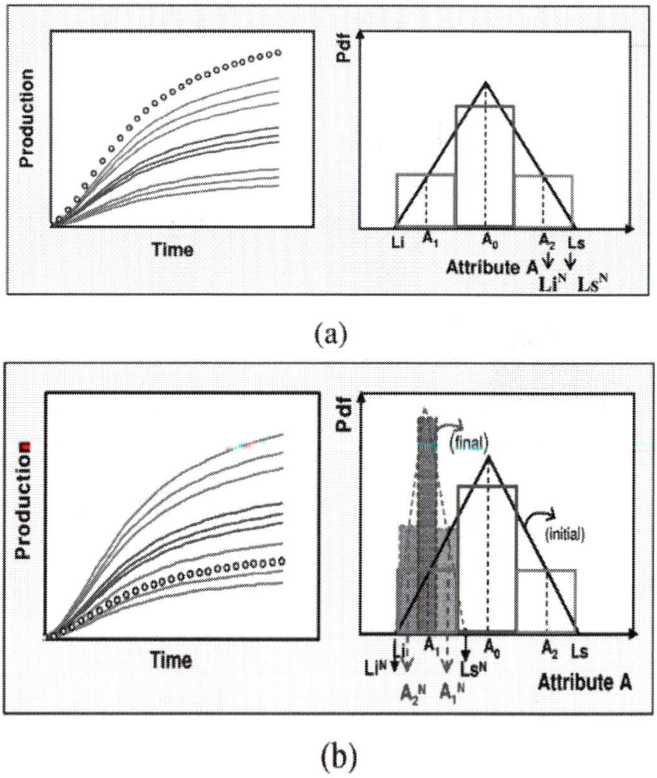

(a)

(b)

Figure 3: Method 3: redefinition of attribute limits.

And

$$Li^{\cdot N} = A_2$$

(11)

The new probable level is calculated as:

$$A_0^N = (Ls^N + Li^{\cdot N}) / 2$$

(12)

Considering the example shown in Fig. 3(b), the new probable level is calculated as follow:

$$A_0^N = \frac{A_0 \times P[A_0] \times \left(\dfrac{1}{|S_0|}\right) + A_1 \times P[A_1] \times \left(\dfrac{1}{|S_1|}\right)}{P[A_0] \times \left(\dfrac{1}{|S_0|}\right) + P[A_1] \times \left(\dfrac{1}{|S_1|}\right)}$$

(13)

And the new upper and lower limits are given by:

$$s^N = A_1 + (A_0 - A_1) \times P[A$$

(14)

$$^r = Li + (A_1 - Li) \times (1 - P[$$

(15)

With the calculation of the new limits and most probable level, according to the triangular distribution, the new pessimistic and optimistic levels (A_1^N and A_2^N) are obtained. There are several possible conditions for obtaining the new values of the uncertainty attributes of the reservoir with triangular distribution. The same considerations are valid in the case of adoption of other types of probability distributions (normal, lognormal, and uniform, among others).

The attempt is made to modify the uncertainty curve of the OF being studied, by the application of these methods, in the direction presented by Fig. 1(a) (left down picture), or in other words, bring it closer to the vertical axis representative of the history. Several attempts were made for the calibration and practical application of the methods for a complex model.

Proposed Changes

The following items are improvements proposed to the methodology presented initially by Moura Filho (2006):

Choice of Local Objective-function

It is suggested, for local-level analysis, the combination of variables Q_w (water rate) and P_{wf} (bottom-hole pressure) measured in the wells (Moura Filho used only Q_w). Equation (16) presents the OF used for wells history matching. It contains different factor of relative weight for each variable, as a function of its validity and degree of importance.

$$OF = \sum_{i=1}^{n} \left[wi_{Q_w} \left(Qw^{obsv} - Qw^{calc} \right)^2 + wi_{P_{wf}} \left(Pwf^{obsv} - Pwf^{calc} \right)^2 \right]$$

(16)

Where wiQ_w and wiP_{wf} are weight for water rate and bottom-hole pressure, respectively.

Probabilistic Scenario Treatment

Modifications are made of the original formulation.

Definition of the Target Uncertainty Range

This allows for evaluation as to whether the uncertainty reduction process should be refined.

Case Analysis Before and After Uncertainty Reduction

An evaluation of the integration consistency at this point permits restarting the process at the well or regional level. This indicates the interactive character of the methodology.

New Sensitivity Analysis

Other variables having been discarded originally could influence the OF at this stage. This analysis is made to reinstate the convenience of including additional attributes in the process and re-start a new step.

Uncertainty Reduction Analysis of the Predictions

Calculation of the uncertainty range reduction after application of the proposed methods on the predictions of main variables of the model.

Method Modifications

After applying the original equations, several alterations attempts were made on weights and calculations of new associated probabilities to apply the methodology to a complex case. The weights act on the alteration of probabilities of the uncertainty levels and the variation of attribute values. At the stage of OF global evaluation, the effects of attribute uncertainty reduction act together, cumulating the dislocations on the newly generated uncertainty curve. Figure 4(a) schematizes a situation in which the uncertainty curve obtained after application of Method 1 manifests an undesirable effect caused by an increase in relative uncertainty for positive values of the OF between 0 and 40% approximately. Positive OF values mean that the values calculated are smaller than the observed values.

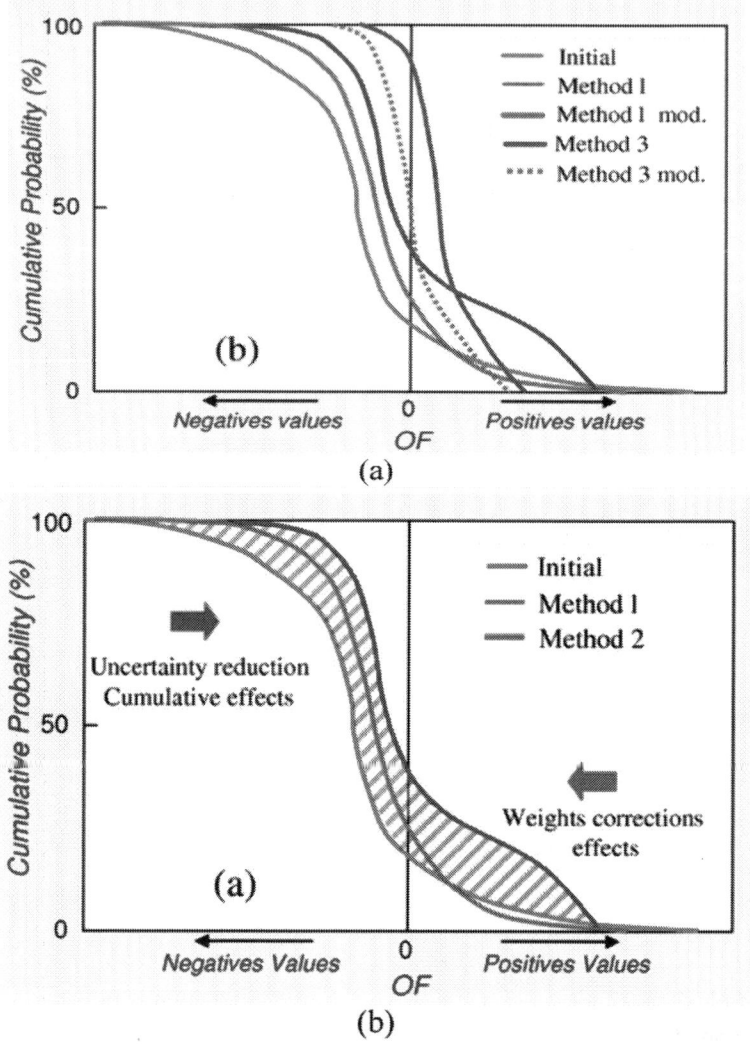

Figure 4: Theoretic uncertainty curves: correction of cumulative effects (a) and comparison of the methods (b).

Thus, it is necessary some revision or variation on the weights assigned to these models, in which calculated curves are below the history values. The selected variation criterion is related to the standard deviation of misfit distances calculated for all the analyzed models, in relation to the chosen OF variable during the history period. Additionally, the models should be arranged according to the pessimistic, optimistic and most probable levels, beginning with the attribute of greatest sensitivity and continuing with the remaining attributes. Thus, for each analyzed attribute, the standard deviation is calculated as:

$$\sigma_n = \sqrt{\frac{\sum\left(x - \overline{x}\right)^2}{Mn}}$$

(17)

Where

$$x = K = \sum_{i=1}^{N}\left(d_i^{obsv} - d_i^{sim}\right)$$

(18)

And

$$\overline{x} = \frac{\sum x}{Mn}$$

(19)

The smallest standard deviation value also corresponds to the attribute with the greatest 1/S value. Next, an F_n factor based on the inverse of standard deviation of each uncertain level permits the modification of initially calculated probabilities P_n. In this manner, a smaller weight is given to the levels originally of greater importance in the combined models with positive OF values.

$$F_n = \frac{\left(\frac{1}{\sigma}\right)_n}{\sum_{i=1}^{n_A}\left(1/\sigma_i\right)}$$

(20)

Where n_A is the number of chosen attributes.

The new probabilities are calculated according to Eq. (21), in which the factor F_n is the proposed change in the original manner of calculating P_n shown in Eq. (4).

$$P_n^{\mathrm{mod}} = P_n \times \frac{1}{F_n}$$

(21)

In Figure 4(a), the effects of the applied correction on the uncertainty curve are also shown. At the global evaluation stage of OF, the effects of attribute uncertainty reduction act in conjunction on the newly generated uncertainty curve.

The improvement of this method also produces a similar effect on the subsequent methods. Thus, Methods 2 and 3 are modified beginning from the use of P_n^{mod} and the new weights are calculated. Modified Method 3 is defined from the parameters obtained in Method 1 corrected, following the explained procedure. The variation of limits is only applied on those attributes having great weight variation, being that it is readily possible to obtain an uncertainty curve that is centered in relation to the OF value of zero, however, slightly more inclined. Figure 4(b) exemplifies the shift of the uncertainty curve from Method 3.

Definition of an Uncertainty Range Target

This range should be based on the value of the objective-function of the curves considered as acceptable limits, selected according to the value of a percentage of the total range between the extreme cases considered. In the example of Fig. 5(a), the maximum negative dispersion is calculated from the difference between the history matching corresponding to the smallest value of the OF with a negative sign (*Min. OF Neg. Matching red curve*) and the history matching that corresponds to the maximum negative OF value (*Maximum OF Neg.*). Thus, the maximum positive range is calculated from the difference between the matching corresponding to the smallest OF value with a positive sign (*Min. OF Pos. OF Matching blue curve*) and the matching corresponding to the maximum negative value acceptance (*Maximum*

OF Positive). From the calculation of these extreme ranges and by means of the choice of an acceptance percentage of each total range for each sign, it is possible to identify the acceptable limits. These limits have OF values of the closest models within an acceptable tolerance limit (*Negative Acceptable Limit and Positive Acceptable Limit*). Figure 5(a) shows an example for the case of a specific percentage choice of the total range. Thus, after the identification of the aforementioned cases, the calculated limits can be plotted on the uncertainty curve graph, permitting the qualitative and quantitative measurement of the degree of uncertainty reduction reached through the application of the methods (Fig. 5(b)).

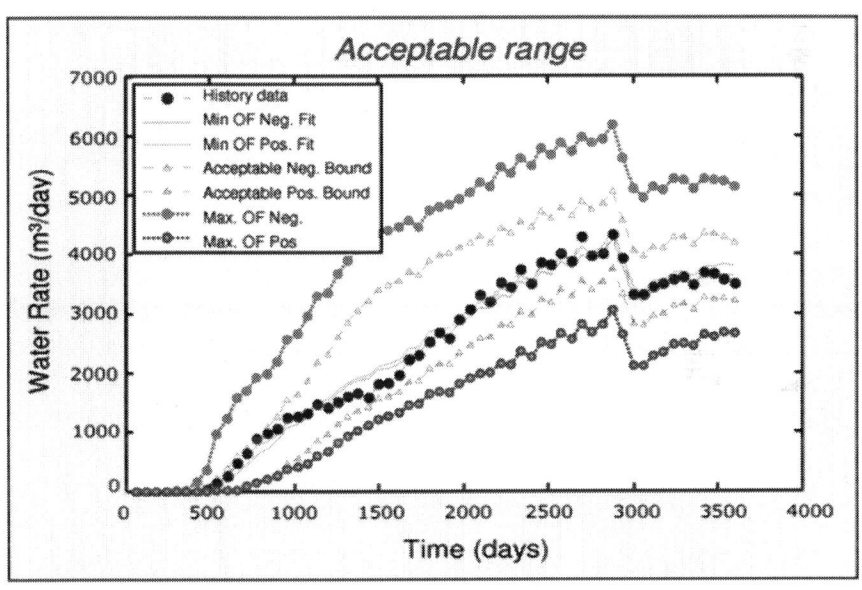

(a)

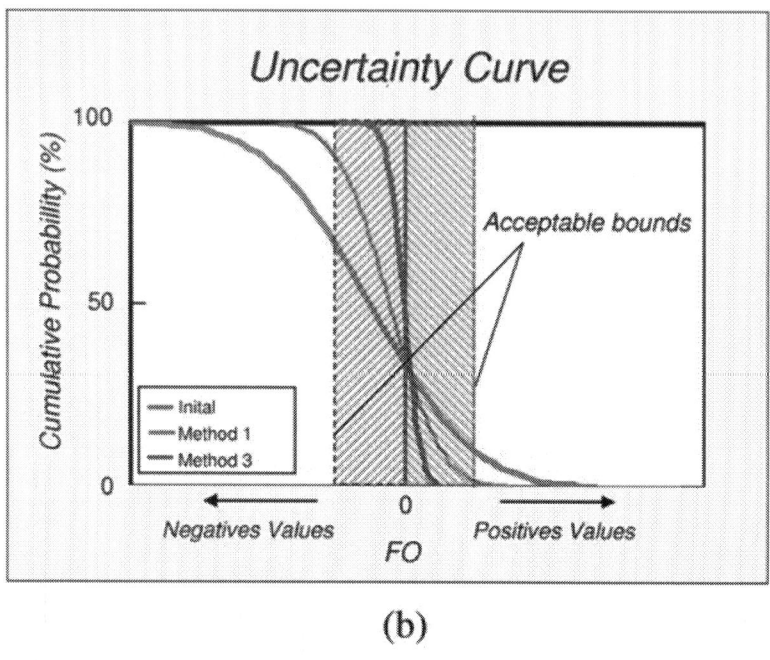

(b)

Figure 5: Definition of an uncertainty range target: selection of bound models (a), target range and uncertainty curves (b).

Integration of Global, Regional and Local Stages

An interactive five phase's process is proposed, in the scope of the reduction uncertainty and evaluation, from this procedure.

Phase 1

Application of the described methods, over the chosen OF with global scope, until obtaining acceptable results (Fig. 6(a)) in this way, through an interactive process, new simulations are performed directed by Method 3 until an acceptable reduction is reached. As a result, a range of curves of global production smaller than the initial dispersion is obtained, and this range is positioned around the observed data.

Phase 2

In this phase, the local stages of history matching integration at the regional and well levels begin. In Fig 6(b) the process is schematized. Matching by zones is performed, proceeding from the choice of the best global matching from the previous phase. In this phase, manual or automated history matching methodologies can be used.

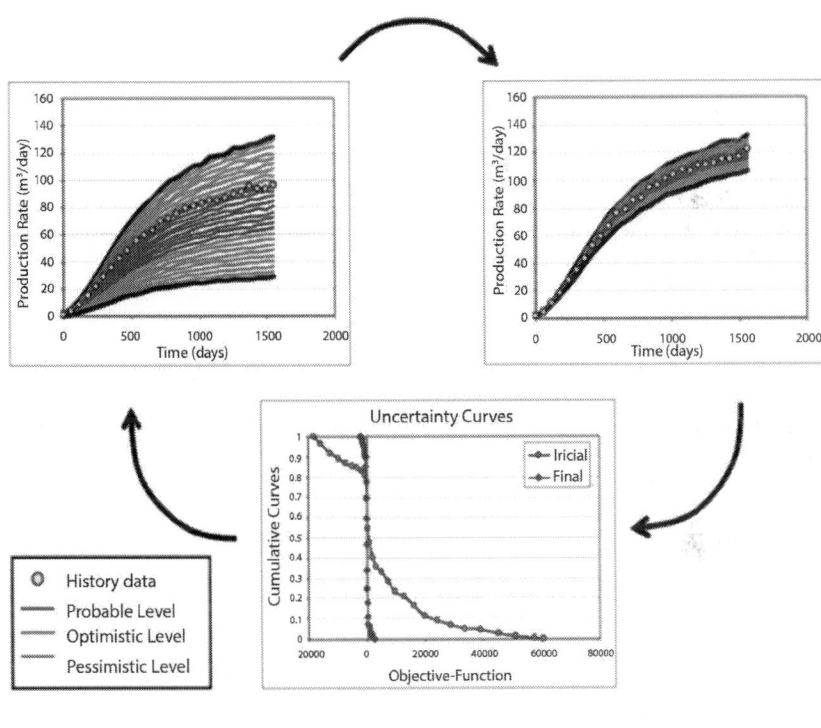

(a)

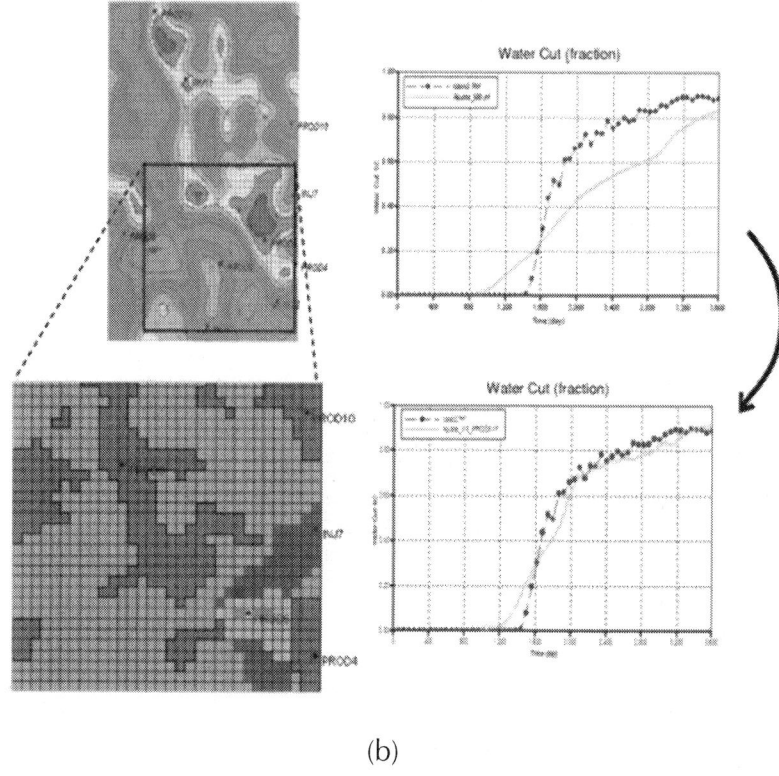

(b)

Figure 6: Integration of global (a) and local analysis (b).

Phase 3

All the modifications at the regional and well levels, explored at the previous stage, are considered. Obtaining new combinations of models, considering the uncertainty still present in the zones where little information is found, permits evaluation of the degree of uncertainty based on the observed data. Figure 7(a) illustrates the final profiles obtained after the reconstruction of the derivative tree with improved local matching.

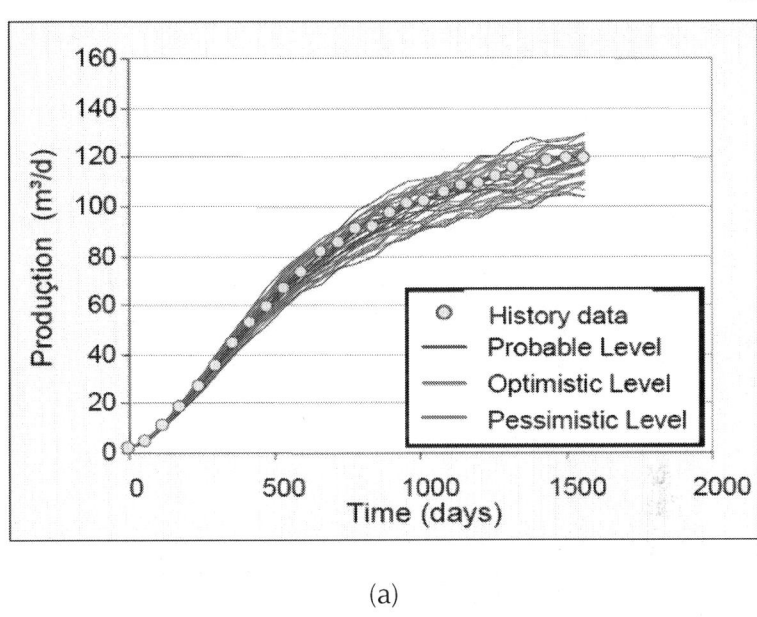

(a)

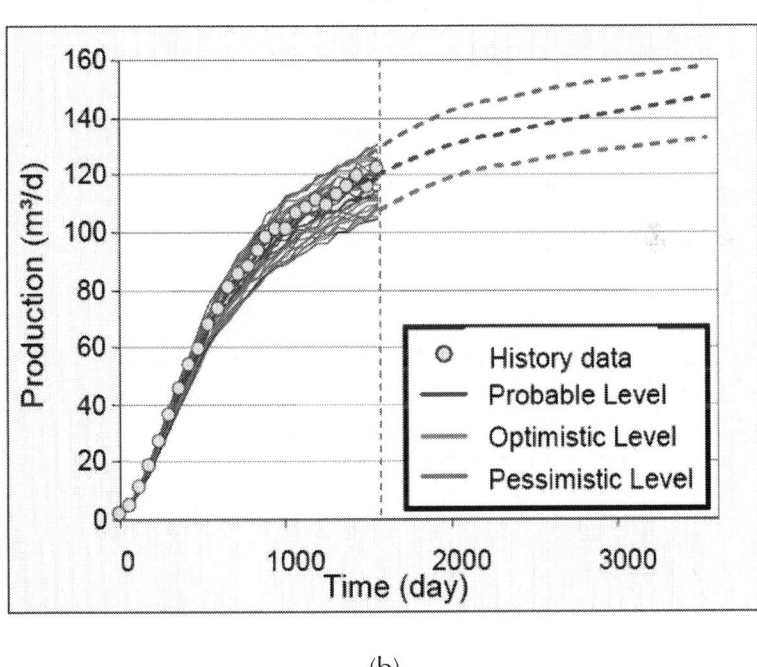

(b)

Figure 7: Schematic final dispersion in history period (a) and in prediction period (b).

Phase 4

It is necessary to keep control of the results obtained to be in accordance with the acceptable limits determined in the beginning of the process. This evaluation phase is critical. If the uncertainty curves obtained in the previous phases are not included in this range, the whole process can be started again, this being the interactive character of the methodology.

Phase 5

A final range of uncertainty of the dynamic performance of the reservoir in the prediction period is reached (Figure 7(b)). The models corresponding to the percentiles 10% and 90% (although other percentiles can be chosen) of the uncertainty curve accepted in the previous phase are appropriate indicators for future performance with uncertainty, after applying the methodology.

Summarizing, the main changes and improvements in the methodology presented originally by Maschio et al. (2005) and Moura Filho (2006) are: 1) Equation (4) was changed (see Eq. (21)); 2) the use of local objective function; 3) the definition of a target uncertainty range; 4) the integration of global and local analysis and 5) the application to a more complex case.

APPLICATION

The methodology was applied in a reservoir model based on outcrop data from Brazil, including well information and seismic interpretations of analog fields in turbidity systems deposited in deep water. The data were treated, qualitatively and quantitatively, for the parameterization of the reservoir. The chosen objective-function is based on monthly water production for the evaluation at global level; however, special attention was given to the well bottom-hole pressures in the phase of local application.

The modeled depositional elements are channels, lateral deposits and hemi-pelagic shales, which represent pauses in the dominant sedimentation process of a turbidite system in deep water. The petrophysical parameters (porosity and permeability) were attributed from correlations with the net-to-gross ratio (NTG), the values being representative of the typical range of existing reservoirs of the Brazilian continental platform (Silva et al., 2005). The refined static model obtained has a grid with 217 x 275 x 6 blocks, with 12 vertical wells, 7 producers and 5 injectors. This model permitted the generation of the synthetic production data taken as reference, for a period of 10 years. This data was subjected to a random noise to represent the common production measurement errors.

Finally, to reproduce the typical conditions of model building in real conditions a second model was constructed from the refined geologic model, to represent the dynamic behavior of the reservoir. The size of the coarse grid model is 43 x 55 x 6 blocks, and with the purpose of changing the original geological conditions, the parameters of the considered elements in each layer were modified following other depositional patterns typical of this environment. Figure 8 shows a three-dimensional view of the corner-point grid used with the spatial distribution of porosity.

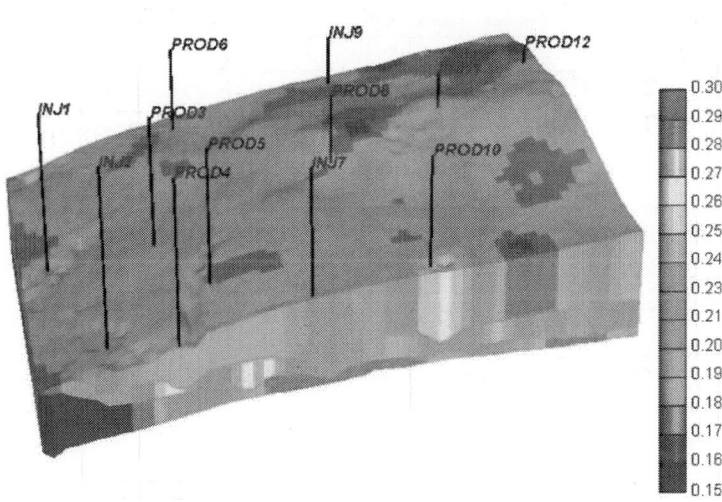

Figure 8: Three-dimensional view of the model studied (porosity).

After the choice of the uncertain static and dynamic attributes, their global and local influence on the model is evaluated. Each uncertain attribute is discretized into three levels with a probability of 20%-60%-20% considering a triangular probability distribution function. Table 1 lists the most probable values and the pessimistic and optimistic levels of the considered attributes. The listed extreme values were used in the sensitivity analysis.

Table 1: Description of the uncertain attributes

Attribute	RepresentativeValues	Description
$NTGf0$	(0.6)	Net to Gross ratio (poor reservoir zone)
$NTGf_1$	(0.2)	
$NTGf_2$	(1.0)	
Ka_0	(2.0)	Absolute permeability multiplier sandstone zone
Ka_1	(1.0)	
Ka_2	(3.4)	
Kf_0	(0.65)	Absolute permeability multiplier (poor reservoir zone)
Kf_1	(0.30)	
Kf_2	(1.00)	
KK_0	(0.12)	Vertical vs. horizontal permeability ratio
KK_1	(0.05)	
KK_2	(0.35)	
Vma_0	(1.20)	Porous volume multiplier of sandstone zone
Vma_1	(0.85)	
Vma_2	(1.55)	
VMf_0	(0.85)	Porous volume multiplier (poor reservoir zone)
VMf_1	(0.61)	
VMf_2	(1.10)	
Kra_0	(0.40)	Relative permeability of the sandstone
Kra_1	(0.26)	
Kra_2	(0.54)	
Krf_0	(0.60)	Relative permeability (poor reservoir zone)
Krf_1	(0.40)	
Krf_2	(0.90)	
BAR_0	(0.50)	Horizontal barriers
BAR_1	(0.00)	
BAR_2	(1.00)	
RIJ_0	(1.25)	Horizontal permeability anisotropy
RIJ_1	(1.00)	
RIJ_2	(1.75)	

BARK$_0$	(0.65)	Vertical seals
BARK$_1$	(0.40)	
BARK$_2$	(0.88)	
PVT$_0$	(790)	Oil density
PVT$_1$	(725)	
PVT$_2$	(855)	

RESULTS

The analysis of the history matching quality is performed in two ways: 1) from the production curves, by observing the reduction of dispersion and comparing it with the observed data and 2) by obtaining the OF's cumulative probabilities curve (uncertainty curve). In this case, the reduction of the uncertainty degree, after application of the methodology, can be evaluated in function of the dispersion around the zero axes, taking into account the target uncertainty range.

Initially, a comparison among the methods presented by Maschio et al. (2005) and Moura Filho (2006) and the modified methods proposed in this paper is depicted in Fig. 10. This figure shows the improvements in the uncertainty curves obtained.

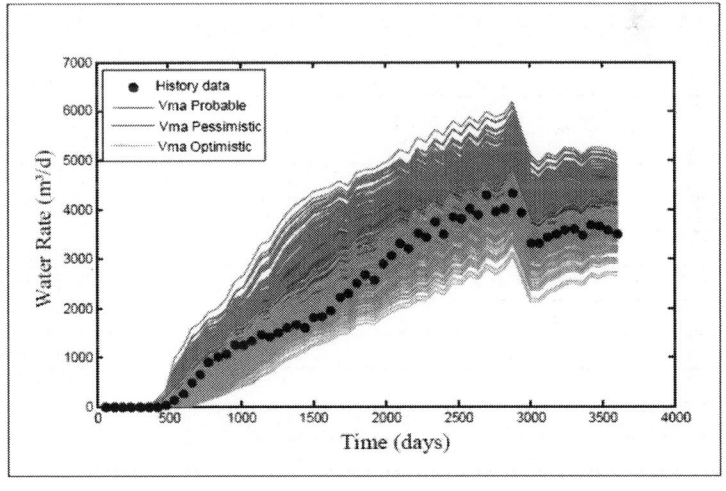

(a)

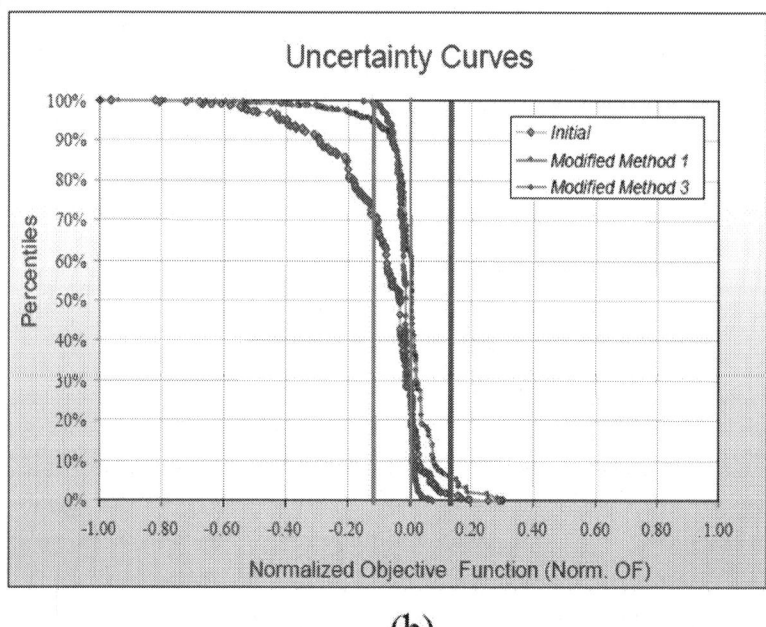

(b)

Figure 9: Probabilistic profiles of total field water production grouped according to V_{ma} (a) and target uncertainty range in the case of 25% of the total range (b).

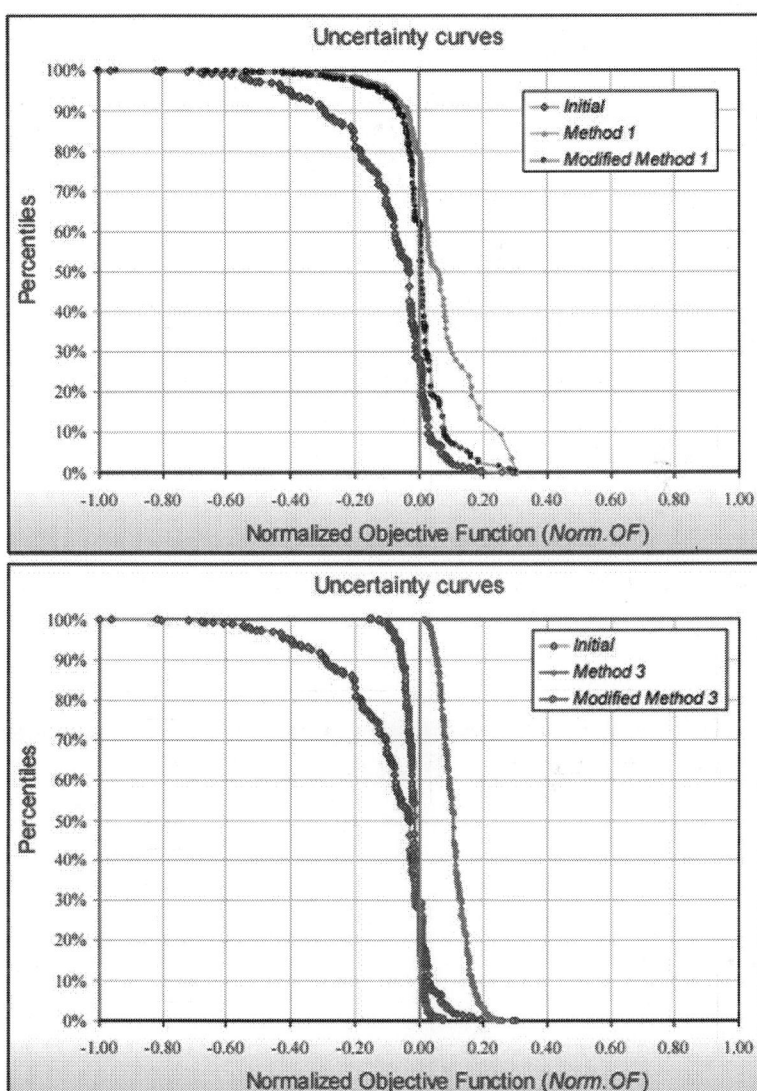

Figure 10: Comparison of Method 1 and 3 proposed by Maschio et al. (2005) and Moura Filho (2006), and Modified Method 1 and 3 (present work).

Additionally, it was also done an analysis of production predictions. From the five critical attributes selected from the sensitivity analysis, discrete in three levels of uncertainty, 35 = 243 simulations were necessary. In Fig 9(a) the water production curves, grouped following the V_{ma} levels (porous volume in the reservoir zone), are presented as an example. Similar curves are made for all the attributes and, to make the process automatic, the curve differences are quantified and used for the probability changes of the attributes.

The new probability value calculated by Method 1 for Vma1 (pessimistic) is 5.4%, for Vma0 (probable) is 11.7% and for Vma2 (optimistic) is 82.9%. It can be seen that the green curves corresponding to the optimistic level of V_{ma} are closer to the recorded data. Figure 9(b) shows the uncertainty curves that indicated the degree of quality of the history matching (the OF shows normalized deviation from the history). The uncertainty curves obtained by the proposed methods demonstrate a significant uncertainty reduction, Method 3 being the most effective. In this figure, an acceptance range of 25% with respect to the interval of initial variation was considered.

Figure 11(a) shows the new disposition of productive profiles obtained from the models constructed after using Method 3 at the global level. The deviation reduction with respect to the history data is expressive, in addition to being well distributed around the observed data for all uncertainty levels.

Field

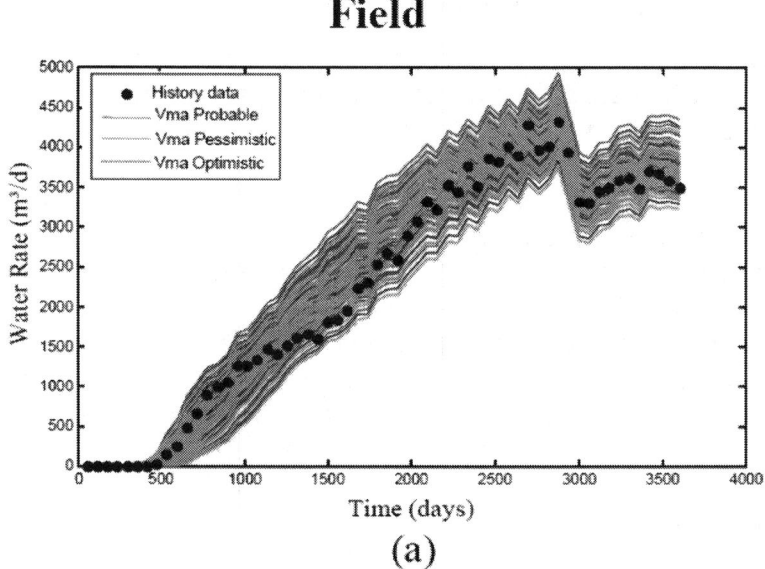

(a)

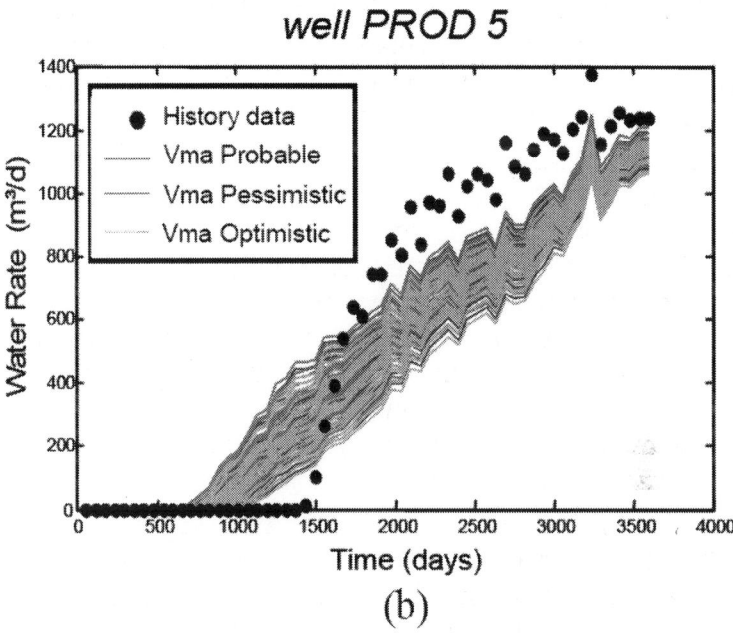

Figure 11: Probabilistic profiles of total field (a) and well PROD5 (b) water production after the application of the methodology at global scale.

Results were also generated for an additional local history matching step integrating the global and local matching processes. Figure 11(b) shows the profiles obtained after Method 3, at a local level, for one well of the model. Great uncertainty reductions in all the wells were obtained with the application over an OF with global scope; nevertheless, as it is evident in the case of well PROD5, the global uncertainty reduction is insufficient to improve the local well matching. This situation shows the necessity of a second stage to correct local matching. Including these data in the analysis permits obtaining probabilistic profiles more centered on the history data of each well, although there continues to exist uncertainty in the model because of the lack of data in the regions between wells or in underdeveloped regions, where sampling is not direct.

Two different approaches were taken. In the first approach, the identification, at the regional level, of the wells with more influential overlapping attributes was proposed, in order to subsequently reinitiate the application of the methodology over this region, permitting the reduction of uncertainty around each well. Finally, in each well's influence area, it is selected a combination of uncertain attributes having lower OF values derived from the application of this method. In the second approach, more traditional, the zones close to the wells are modified locally and individual local matching is made without modifying the zones of the remaining wells. This approach is not covered in this paper.

Different regions were selected by identification of zones with high coincidence of attributes that are more influential over the OF, based on water production and dynamic pressure of the target wells. Then, over each region, the methodology is applied for each well, now with the OF defined in Eq. (16), with weight factors wi_{Qw} and wi_{Pwf} with values of 0.75 and 0.25 respectively. The initial uncertain attributes chosen are K_v (vertical vs. horizontal permeability ratio), K_{rw} (relative sand permeability), K_a (absolute sand permeability multiplier), K_f (multiplier of permeability in the non-reservoir zone) and VMf (porous volume non-reservoir zone).

The variation range of some of these attributes was already reduced in the global treatment of the previous phase. In the case of well PROD3, Fig. 12(a) shows the initial spread of the curves for water rate in reference to the 243 simulation models. In Fig. 12(b), the curve distribution shown refers to the models matched after application of Method 3 modified for the well under analysis. An important narrowing of the range of history matching is verified as expected.

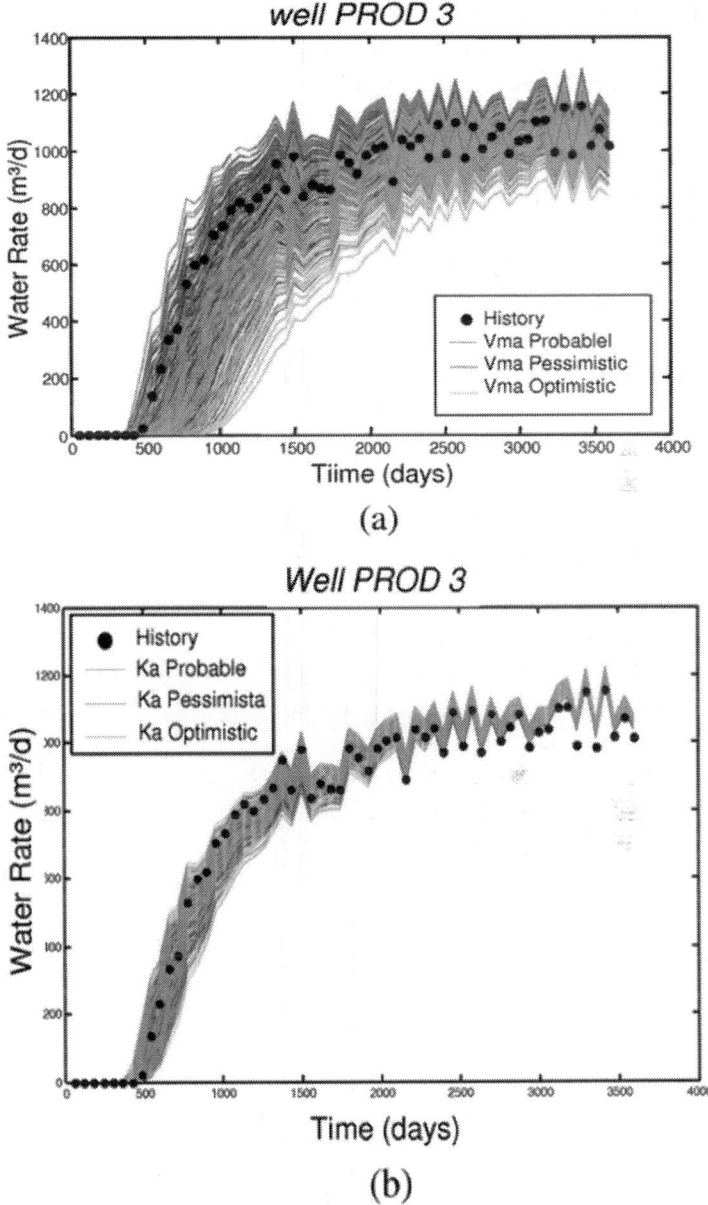

Figure 12: Initial probabilistic profiles of water rate (a) and after the application of the methodology at local scale (b).

The local history matching obtained in the previous phase for all selected regions were combined together in the base case considering the reduced uncertainty ranges in Phase 1 for the rest of the reservoir. Finally, with the same limits identified with modified Method 3, the possible combined models are obtained. Figure 13(a) presents the final distribution of the probabilistic profiles of total water production. The obtained set of curves, with less dispersion and well centered in relation to the observed values, represents the final solution. The next phase is the result control through the definition of target ranges (or acceptable limits) for the process of uncertainty reduction. The demarcation limits of the target range are shown in Fig. 13(b). The chosen acceptance range, in this case, is 25% of the total spread. In Fig. 14(a), there are plotted the corresponding curves for acceptable limits.

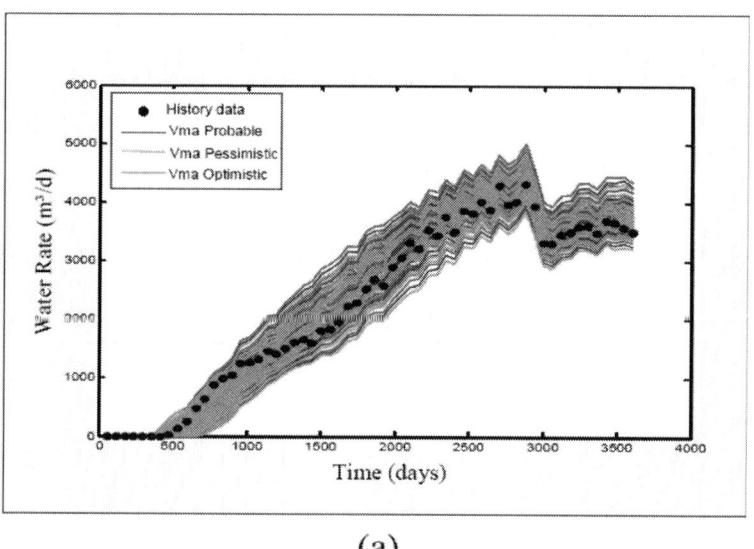

(a)

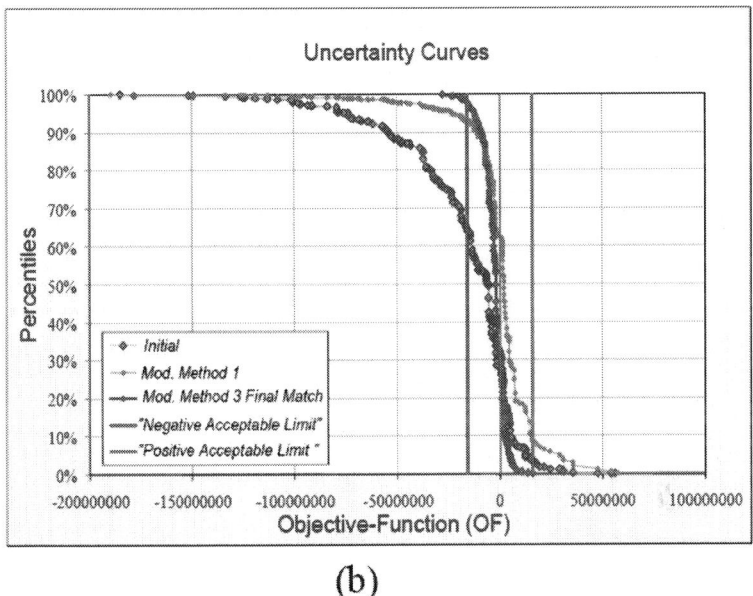

(b)

Figure 13: Final probabilistic profiles (a) and uncertainty curves (b).

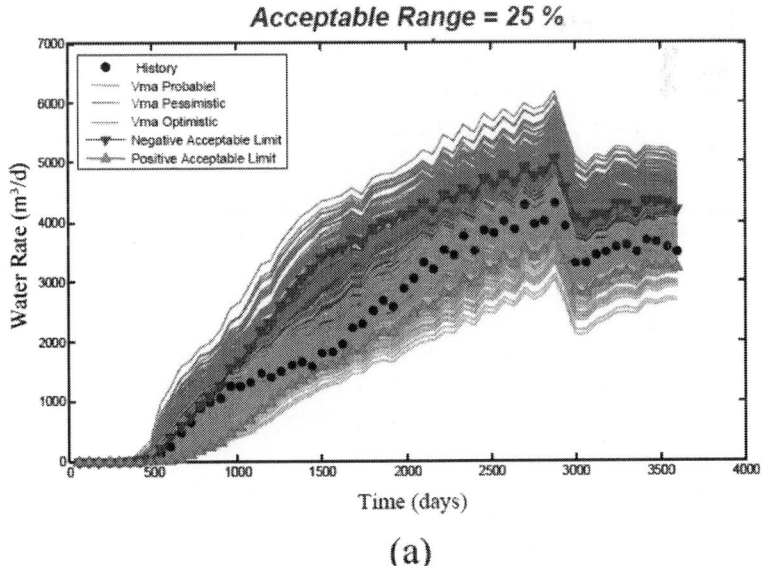

(a)

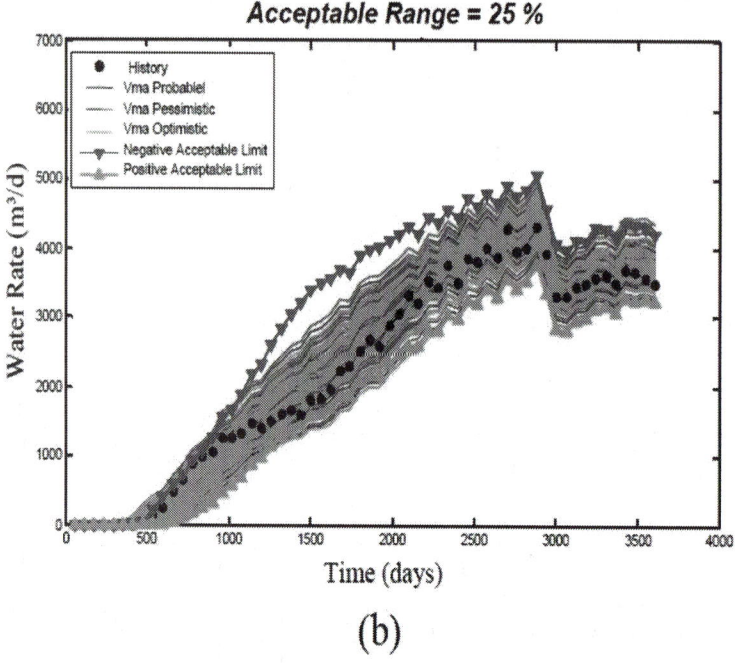

(b)

Figure 14: Acceptable limits versus initial dispersion of probabilistic profiles (a) and acceptable limits versus final dispersion of probabilistic profiles (b).

The uncertainty curve constructed at the end of Phase 3 (Fig. 13(b)) fits, almost entirely, within acceptable limits, demonstrating that the process reached its objectives. This can be confirmed in Fig. 14(b), by contrasting the chosen limits to the final dispersion of the probabilistic profiles after Method 3.

The reduced ranges of the critical attributes allow to a consequent reduction of production prediction spread. The prediction of the water rate of the models representing the percentiles P10 and P90 is reported in Fig. 15. These models were chosen from the uncertainty curve before (Fig. 15(a)) and after (Fig. 15(b)) the application of the methodology presented in this work.

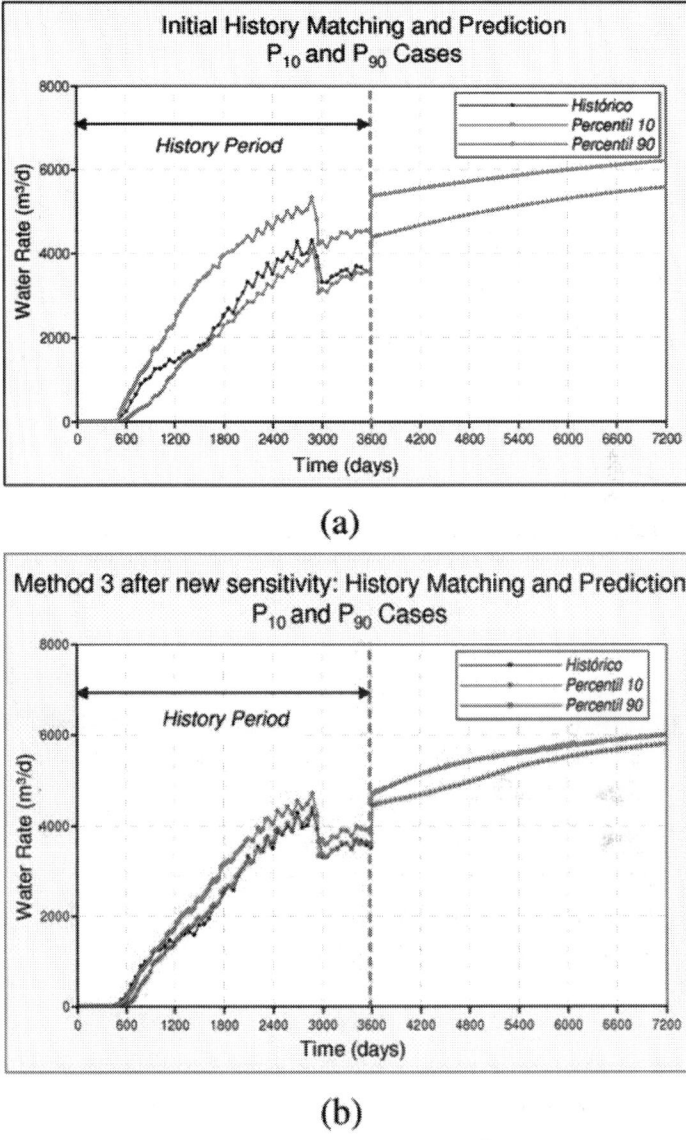

(a)

(b)

Figure 15: Prediction of the models P10 and P90 after (a) and before (b) the application of the methodology.

Finally, Figs. 16(a) and 16(b) present the values obtained from accumulated oil production (millions of m3) and water production (millions of m3) for each percentile and according to the applied phase. It can be clearly seen a gradual reduction of the difference between P10 and P90.

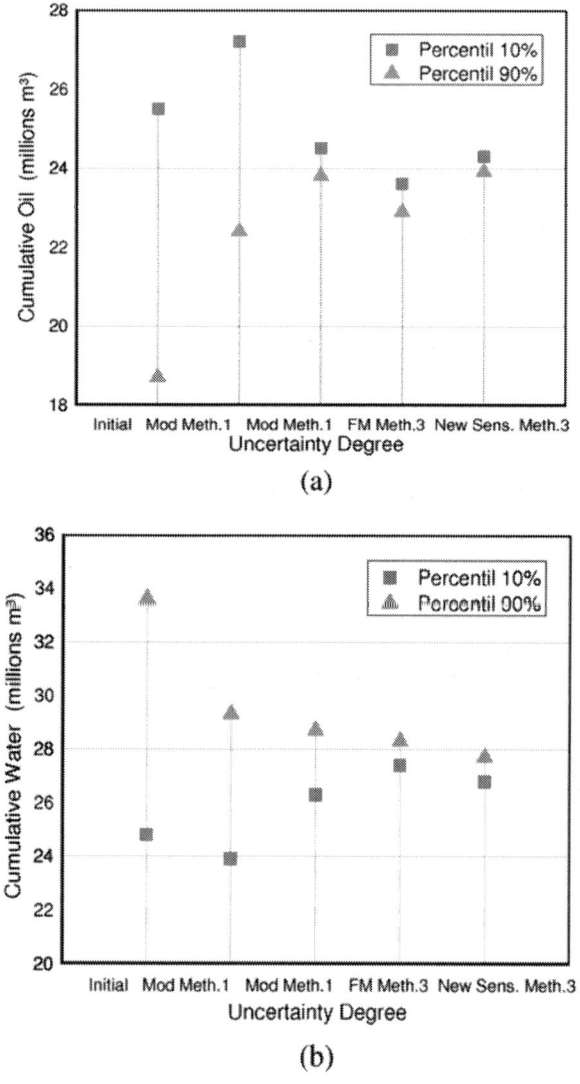

Figure 16: Cumulative oil (a) and water (b) as a function of degree of uncertainty.

CONCLUSIONS

A consistent and flexible methodology to integrate history matching with uncertainty analysis at global, regional and local levels in a complex model was presented in this paper. The application allows obtaining the following additional conclusions:

- The used methods permitted: 1) reduction of the range of possible history matching; 2) identification and conditioning of the uncertainty present as function of the observed data; 3) reduction of the uncertainty intervals of the identified critical attributes; and 4) demarcation of confident limits for the reservoir's future performance.

- The focus on how to approach the history matching, when there is a set of highly variable attributes and restricted knowledge, was changed, obtaining a defined group of models that comprise the possible matching with their associated probabilities.

- The sensitivity analyses permitted the detection of uncertain attributes critical to the evaluation of the degree of subsequent uncertainty, thus simplifying the problem as well as reducing significantly the number of attributes and, consequently, the run time.

- Methods 1 and 2 were faster, as they did not require new simulations. A new calibration of Method 1 was necessary. Method 3 provided greater uncertainty reduction, yet required greater computational effort, compared to Methods 1 and 2.

- The reduction of global uncertainty did not guarantee a local uncertainty reduction. Consequently, it was necessary to take into account the interaction between regions. The applied methodology permitted analysis by stages, which gives great flexibility to application in practical cases.

- Obtaining representative prediction curves for the reservoir (percentiles P10 and P90, for example) permitted an estimation of the risk reduction of the considered project performance.

- The probabilistic approach of the history matching made available a broader vision, as it points to several possible scenarios in the search for the reservoir's real behavior. Nevertheless, the final choice of representative scenarios depends on the criteria adopted by the analyst.

- When new data are added to the study, the history matching and the predictions can be improved reducing the attribute range by the application of the complete proposed flow char

Considering all the above listed items, the consequence is increased confidence in the use of the simulation as an auxiliary tool in the decision process. One advantage is flexibility as to the use of different uncertainty analysis tools and the definition of distinct types of probability distribution in order to mark the levels of the uncertain attributes. Another advantage, compared to automated processes of model calibration, is to make unnecessary the use of sophisticated optimization methods. Equipment with parallel processing and software integration makes it possible its application to real cases.

Other methodologies have analogous conclusions or are similar in some points covered. In this paper, a general procedure attempts a progressive mitigation of uncertainty in all phases of a project, incorporating history matching of the model.

The choice of uncertain attributes levels and their variation limits is a crucial step in the process and has to reflect the real uncertainties of the problem. The experience of a multi-disciplinary team is critical at the beginning of the process, as long as in search for representative attributes levels, processed data from analogical basins and fields can be very usetul.

ACKNOWLEDGEMENTS

The authors gratefully thank the Brazilian council for research and development – Conselho Nacional de Desenvolvimento Científico e Tecnológico (CNPq/PROSET), CEPETRO and PETROBRAS (SIGER) for supporting this research. We also thank Petrobras Energia S.A. (PESA) for their financial support.

REFERENCES

1. Alvarado, M.G., McVay, D.A., Lee, W.J., 2005, "Quantification of Uncertainty by Combining Forecasting With History Matching", *Journal of Petroleum Science and Technology*, 23 (3-4), 445-462.

2. Becerra, G.G., 2007, "Uncertainty mitigation through the integration with production history matching", Department of Petroleum Engineering – State University of Campinas, Unicamp. São Paulo, Brazil, 192 p. (Master Sciences, In Portuguese).

3. Bennett, F. and Graf, T., 2000, "Use of Geostatistical Modeling and Automatic History Matching to Estimate Production Forecast Uncertainty - A Case Study", SPE 74389, International Petroleum Conference and Exhibition, Mexico, 10-12 February.

4. Bissel, R.C., 1997, "Combining Geostatistical Modeling with Gradient Information for History Matching: the Pilot Point Method", SPE 38730.Annual Technical Conference and Exhibition, San Antonio, Texas, U.S.A., 5-8 October.

5. Christie, M., MacBeth C., Subbey S., 2002. "Multiple history-matched models for Teal South", The Leading Edge, 21 (3), 286-289.

6. Gu, Y. and Oliver, D.S., 2004, "History Matching of the PUNQ-S3 Reservoir Model Using the Ensemble Kalman Filter", SPE 89942. Annual Technical Conference and Exhibition, Houston, Texas, U.S.A., 26-29 September.

7. Guérillot, D.and Pianelo, L., 2000, "Simultaneous Matching of Production Data and Seismic Data for Reducing Uncertainty in Production Forecasts", SPE 65131.European Petroleum Conference, Paris, France, 24-25 October.

8. Jenni, S., Hu, L.Y., Basquet, R., de Marsily, G. and Bourbiaux., 2004, "History Matching of Stochastic Models of Field-Scale Fractures: Methodology and Case Study", SPE 90020, Annual Technical Conference and Exhibition, Houston, Texas, U.S.A., 26-29 September.

9. Kashib, T., Srinivasan, S., 2006, "A Probabilistic Approach to Integrating Dynamic Data in Reservoir Models",Journal of Petroleum Science and Engineering, 50 (3-4), 241-257.

10. Landa, J.L. and Guyaguler, B., 2003, "A Methodology for History Matching and the Assessment of Uncertainties Associated with Flow Prediction", SPE 84465, Annual Technical Conference and Exhibition, Denver, Colorado, U.S.A., 5-8 October.

11. Lépine, O.J, Bissel, R.C., Aanonsen, S.I, Pallister, I.C., Barker, J.W., 1999, "Uncertainty Analysis in Predictive Reservoir Simulation

Using Gradient Information", SPE 57594, *SPE Journal*, 4 (3), 251-259.

12. Litvak, M., Christie, M., Johnson, D., Colbert, J. and Sambridge, M., 2005, "Uncertainty Estimation in Production Constrained by Production History and Time-Lapse Seismic in a GOM Oil Field", SPE 93146, Reservoir Simulation Symposium, Houston, Texas, 31 January-02 February.

13. Ma, X., Al-Harbi, M., Datta-Gupta, A., Efendiev, Y., 2006, "A multistage sampling method for rapid quantification of uncertainty in history matching geological models", SPE 102476, Annual Technical Conference and Exhibition, San Antonio, Texas, October.

14. Manceau, E., Mezghani, M., Zabalza-Mezghani, I. and Roggero, F., 2001, "Combination of Experimental Design and Joint Modeling Methods for Quantifying the Risk Associated whit Deterministic and Stochastic Uncertainties – An Integrated Test Study", SPE 71620. Annual Technical Conference and Exhibition, New Orleans, Louisiana, 30 September-3 October.

15. Maschio, C., Schiozer, D.J. and Moura Filho M.A.B., 2005, "A Methodology to Quantify the Impact of Uncertainties in the History Matching Process and in the Production Forecast", SPE 96613, Annual Technical Conference and Exhibition, Dallas, Texas, 9-12 October.

16. Moura Filho, M.A.B., 2006, "Integration of Uncertainty Analysis and History Matching Process", Department of Petroleum Engineering – State University of Campinas, Unicamp Campinas, São Paulo, Brazil, 150 p. (Master Sciences, In Portuguese)

17. Nicotra, G., Godi, A., Cominelli, A. and Christie, M., 2005, "Production Data and Uncertainty Quantification: A Real Case Study", SPE 93280, Reservoir Simulation Symposium, Houston, Texas, 31 January-02 February.

18. Queipo, N.V., Pintos, S., Rincón, N., Contreras, N., 2002, "Surrogate Modeling-Based Optimization for the Integration of Static and Dynamic Data into a Reservoir Description", *Journal of Petroleum Science and Engineering*, 35 (3-4), pp. 167-181.

19. Reis, L.C., 2006, "Risk Analysis with History Matching using Experimental Design or Artificial Neural Networks", SPE 100255,

Europec/EAGE Annual Conference and Exhibition, Vienna, Austria, 12-15 June.

20. Roggero, F., 1997, "Direct Selection of Stochastic Model Realizations Constrained to Historical Data", SPE 38731, Annual Technical Conference and Exhibition, San Antonio, Texas, 5-8 October.

21. Rotondi, M., Nicotra, G., Godi, A. et al., 2006, "Hydrocarbon Production Forecast and Uncertainty Quantification: A Field Application", SPE 102135, Annual Technical Conference and Exhibition, San Antonio, Texas, 24-27 September.

22. Schiozer, D.J., Almeida Netto, S.L., Ligero, E.L., Maschio, C., 2005, "Integration of History Matching and Uncertainty Analysis", *Journal of Canadian Petroleum Technology*, Vol 44, No.7, pp 41-46, July.

23. Silva, F.P.T., Rodríguez, J.R.P., Paraizo, P.L.B., Romeu, R.K., Peres, A.M.M., Oliveira, R.M., Pinto, L.B. and Maschio, C., 2005, "Novel Ways of Parameterizing the History Matching Problem", SPE 94875, Latin American and Caribbean Petroleum Engineering Conference, Rio de Janeiro, 20-23 June.

24. Suzuki, S. and Caers, J., 2006, "History Matching With an Uncertain Geological Scenario", SPE 102154, Annual Technical Conference and Exhibition, San Antonio, Texas, 24-27 September

25. Varela, O.J., Torres-Verdín, C., Lake, L.W., 2006, "On the Value of 3D Seismic Amplitude Data to Reduce Uncertainty in the Forecast of Reservoir Production", *Journal of Petroleum Science and Engineering*, 50 (3-4), pp. 269-284

26. Williams, G.J.J., Mansfield, M., MacDonald, D.G., Bush, M.D., 2004, "Top-Down Reservoir Modelling", SPE 89974, Annual Technical Conference and Exhibition, Houston, Texas, SPE, 26-29 September.

27. Zabalza-Mezghani, I., Manceau, E., Feraille, M., Jourdan, A., 2004, "A. Uncertainty Management: from Geological Scenarios to Production Scheme Optimization", *Journal of Petroleum Science and Engineering*, 44 (1-2), 11-25.

Chapter 4

2D Model Study of CO_2 Plumes in Saline Reservoirs by Borehole Resistivity Tomography

Said A. al Hagrey

Department of Geophysics, University of Kiel, Otto-Hahn-Platz 1, 24118 Kiel, Germany

ABSTRACT

The performance of electrical resistivity tomography (ERT) in boreholes is studied numerically regarding changes induced by CO_2 sequestration in deep saline reservoirs. The new optimization approach is applied to generate an optimized data set of only 4% of the comprehensive set but of almost similar best possible resolution. Diverse electrode configurations (mainly tripotential α and β) are investigated with current flows and potential measurements in different directions. An extensive 2.5D modeling (>100,000 models) is conducted systematically as a function of multiparameters related to hydrogeology, CO_2 plume, data acquisition and methodology. ERT techniques generally are capable to resolve storage targets (CO_2 plume, saline host reservoir, and impermeable cap rock), however with the common smearing effects and artefacts. Reconstructed tomograms show that the optimized and multiply oriented configurations have a better-spatial resolution than the lateral arrays with splitting of potential and current electrode pairs between boreholes. The later arrays are also more susceptible to telluric noise but have a lower level of measurement errors. The resolution advance of optimized and multiply oriented configurations is confirmed by lower values for ROI (region of index) and residual (relative model difference). The technique acceptably resolves targets with an aspect ratio down to 0.5.

INTRODUCTION

The need to manage the global CO_2 emissions for mitigating the greenhouse effect has led to a world wide research to reduce atmospheric CO_2. Techniques of carbon capture and storage (CCS) must (1) be effective and cost-competitive, (2) provide stable, long-term storage, and (3) be environmentally benign. Potential terrestrial media for CO_2 storage include depleted oil and gas reservoirs, unmineable coal seams, and deep saline water reservoirs capped by impermeable rock to prevent upward leakage.

CO_2 exists in the gas phase at standard atmospheric temperature and pressure. Above the dynamic critical point (>31.1°C, >7.38 MP, density >0.469 g/cm³), CO_2 changes to a supercritical fluid phase; it diffuses through solids like a gas and dissolves material like a liquid. CO_2 has long been injected in the subsurface to enhance oil, gas, and coal-bed methane recovery and storage. This injection has been mainly monitored using seismic time-lapse imaging (e.g., Sleipner oil field in North Sea, e.g., [1]). However, investigations on brine-saturated sandstones showed that the electrical resistivity (ρ) is more sensitive to CO_2 saturation than is seismic velocity (Figure 1). This may justify application of electrical resistivity tomography (ERT), particularly in boreholes, for monitoring resistive supercritical CO_2 plumes in a deep saline reservoir (e.g., [2]). This reservoir formation normally consists of a highly resistive matrix (e.g., sandstone and limestone) and a conductive pore brine. Here, CO$_2$ saturation can be predicted using the law of Archie [3]:

$$\rho = a\rho_w \Phi^{-m} S_w^{-n'},$$

(1)

$$\rho_{CO_2} = a\rho_w \Phi^{-m} \left(1 - S_{CO_2}\right)^{-n'},$$

(2)

Where ρ, ρ_w, ρ_{co_2} = bulk, fluid, and CO$_2$ resistivity, respectively, Φ = porosity, S_w, S_{CO_2}, = water, and CO$_2$ saturation, a, m, n = constants.

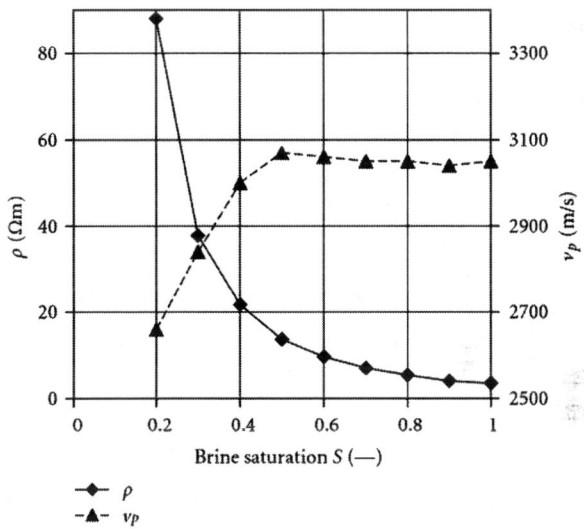

Figure 1: Experimental P-wave velocity (v$_p$) and electrical resistivity (ρ) of sandstone reservoir versus brine saturation (S) showing ρ far more sensitive to S than v$_p$ [4].

Recent developments have enabled installing deep boreholes with coated (insulating) casing and fixed electrode arrays for ERT monitoring (e.g., [5]). Forward modeling and inversion algorithms have also been developed for better monitoring CO$_2$ plume scenarios in deep saline reservoirs and coal seams (e.g., [6, 7]). In 2008 we started the interdisciplinary project "CO$_2$ MoPa" (modeling and parameterization of CO$_2$ storage in deep saline formations for dimensions and risk analysis). It aims at studying long-term CO_2 attenuation and migration in deep and shallow layers (including saline and fresh water aquifers), along with assessing storage capacity and analyzing risk. Various synthetic, almost realistic, storage scenarios are simulated for formations of the North German Basin that seem suitable for CO_2 storage. Our main task is to develop optimized, constrained monitoring strategy techniques for CCS using a combined seismic and ERT approach. This approach focuses on (1) developing a constrained electrical resistivity-modeling strategy based on a priori subsurface knowledge from seismic time-lapse imaging (some years) and logging data, and (2) reliably inverting continuous ERT time-lapse imaging to yield spatiotemporal developments in the intrinsic physico-chemical properties of CO_2 reservoirs and cap rocks with time.

Optimized and Reliable Borehole ERT

Inverse ERT algorithms, however, tend to smear resistivity values from any given voxel to adjacent voxels (e.g., [8]). Based on the sensitivity functions, ERT in boreholes performs near the electrodes (boreholes) better than in the interwell region. Electrical sensitivity is used to select an array for a certain target but one that does not necessarily has the best possible resolution. Thus, a new approach of array optimization was recently developed to search for electrode configurations that maximize survey resolution (e.g., [9]). The optimization algorithms take into account the trade-off between the spatial and temporal (measurement time) resolution. They select measurements based on the contribution to the cumulative sensitivity of the array (e.g., [10]) or the model resolution matrix, R (e.g., [11]). R depends on sensitivities of all configurations plus regularization types used in the inversion [12]. For an arbitrary electrode array, the algorithms generate optimized (opt) data sets that have far less size than the comprehensive one and almost the same resolution of the targets. This comprehensive set includes all possible viable electrode configurations conducted within this array and possesses the maximum possible resolution, see next sections. The application of this 2D array optimization was recently extended into borehole-borehole and surface-borehole surveys [13–15]. The last algorithm is applied here. This algorithm strongly improves the ERT resolution in the interwell region (commonly low) to approach the resolution of the highly sensitive region close the boreholes.

The sensitivity accounts only for data sampling and model heterogeneities. As an alternative, the region of investigation index (ROI) is used to assess the whole 2D inversion procedure such as the data sampling and noise, model discretization and regularization, and nonlinearity [16, 17]. Thus, the reliability of ERT 2D tomograms will be evaluated here by ROI in addition to the relative model difference (residual) between each input and inverted output model.

For ERT in borehole surveys, the topographical aspect ratio (AR) is defined by the vertical length of the electrode array divided by the horizontal crosshole offset. Thus, resolution is enhanced by increasing the density (i.e., number and thus costs) of expensive monitoring wells in the area of plume migration. Newmark et al. [18] studied AR values of 2, 1.5, and 1 and found that they (in this order) show the best, intermediate, and worst resolution, respectively. In this study the lower boundary of AR is extended down to 0.25 which leads to a further decrease in the number (and thus costs) of monitoring wells. A broad AR range (0.25–2) is tested to determine its optimum value between the highest and lowest resolution (of AR = 2 and 0.25, resp.) that corresponds to the highest and lowest number of monitoring wells (i.e., costs), respectively.

Problem and Objectives

Until now ERT is rarely applied for the CCS problematic in deep saline reservoirs. Only few recent studies partly treated this problem, for example, feasibility studies by Christensen et al. [19] and sensitivity investigations for some specified CO_2 plume forms using point and long (metal-cased) borehole electrodes by Ramirez et al. [20]. Also there is a deficit of field sites for CO_2 sequestration that are equipped by adequate acquisition infrastructures for ERT surveys. All these lead to a strong demand for systematic ERT modeling investigations. In this study, extensive, systematic numerical ERT 2.5D modeling is carried out, and the results are analyzed for different virtual scenarios of injected wedge-like CO_2 plumes (dimensions, S_{CO_2} or ρ) as a function of electrode configuration, burial depth, AR, data noise, and setup parameters of modeling constraints (mainly regularization parameters, see next sections). Moreover, ROI analyses and residuals are applied to evaluate the resolving capability of various electrode configurations and inversion procedures. The technique's robustness in the field is tested by adding three different random errors to data sets. These studies aim to test the capability of (non-)standard and optimized ERT techniques (partly developed here) to resolve the subsurface CO_2 storage targets as a function of diverse parameters related to hydro-/geologic and geochemical subsurface properties (mainly of saline reservoir and cap rock), CO_2 plume, survey design, data acquisition, and modeling techniques.

In the next sections, I describe the applied borehole electrode configurations, the experiment setup of the subsurface model scenarios, and modeling varieties. I then discuss, summarize, and conclude the results of the different numerical simulations. Optimization algorithms, noises, depth effect, and modeling constraints are not shown here. They are contained in Hagrey [2, 14, 15] and Hagrey and Petersen [21].

BOREHOLE ELECTRODE CONFIGURATIONS

Similar to surface surveys, ERT data acquisition between two borehole electrode arrays can be conducted in the tripotential 4-pole configurations α (CPPC, C = current electrode, P = potential electrode), β (CCPP) and γ (CPCP), and their reciprocals. The γ configurations can be derived from α and β measurements, that is, they are not independent and are usually excluded from the data [22]. The rest of α and β measurements are accomplished in vertical (v, at 90°), horizontal (h, 0°), and lateral (l, >0–<90°) modes (Figure 2). These 4-pole modes are carried out within the same borehole (inhole) or distributed between the two boreholes (crosshole). Table 1 shows all possible configurations α and β for the inhole and crosshole modes [23]. Each configuration consists of two inhole and three crosshole arrangements.

Table 1: All possible 4-pole tripotential configurations α and β (non-equivalent, nonreciprocal) of a survey between two borehole electrode arrays

Group					
Configure	**Inhole**		**Crosshole**		
	4-0	**0-4**	**3-1**	**1-3**	**2-2**
α	CPPC-0	0-CPPC	CPP-C	C-PPC	CP-PC
β	CCPP-0	0-CCPP	CCP-P	C-CPP	CC-PP

For example, group 4-0 denotes for number of electrodes (current, C/potential, P) in the first (4) and second (0) boreholes, respectively.

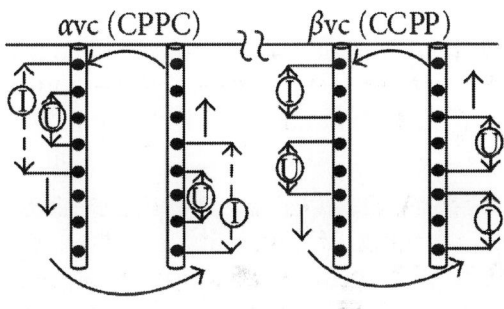

(a)

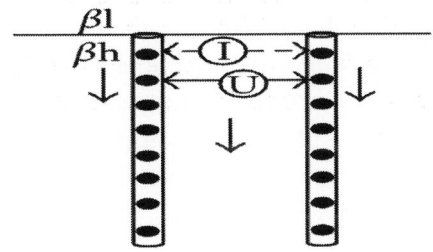

(b)

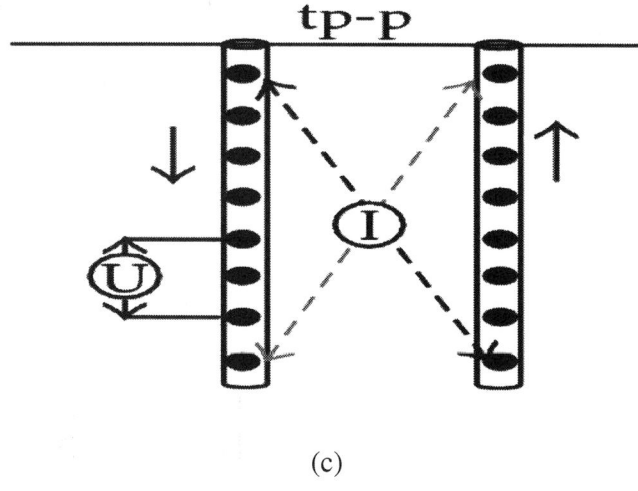

(c)

Figure 2: Tripotential electrode configurations α and β applied for electric resistivity tomography in boreholes (inhole and crosshole) using 4-pole of current (C) and potential (P) electrode pairs. The survey can be conducted in circulating (c) vertical, v (a), lateral, I and horizontal, h (b) modes and tripole-pole, tp-p (c) which is a special type of circulating vertical configuration with fixed C electrodes [24].

For an array of N electrodes, the whole comprehensive data set contains $[N(N - 1)(N - 2)(N - 3)/8]$ independent nonreciprocal 4-pole configurations (Table 2, [25]). Excluding the less stable inversion configurations from this whole set results in a more effective data set, simply called here comprehensive data set. These redundant configurations include γ configurations and those with geometric factors larger than that produced by the dipole-dipole array where the maximum dipole separation is 6a (a = electrode spacing, Figure2). The comprehensive data set with all viable configurations should provide the best possible resolution [26]. It contains all subsurface information that can be gathered by an N-electrode array. Table 2 displays the size of the whole and effective comprehensive data sets in comparison with the standard ones for a collinear 32 electrode array and circulating arrays between two boreholes, each of 16 electrodes. The effective comprehensive set in boreholes contains more than 80,000 data points and is even about 0.65 of its whole set but more than 450 times each of the standard dipole-dipole and Wenner sets. This justifies applying the new approach of electrode optimization to generate optimized data sets of far lower size and almost the same resolution as the comprehensive ones, that is, of highly spatiotemporal resolution (for more details, see [14,15]). Briefly this opt array improves the ERT resolution particularly in the interwell region which is commonly low compared to that of the region close the boreholes.

Table 2: Data set size of 4-pole configurations for a collinear N (32) electrode array and circulating crosshole array of 16 electrodes in each borehole

Data set	Relation	Collinear ($N = 32$)	Circulating crosshole array ($N = 16 \leftrightarrow 16$)
Comprehensive, whole (independent, nonreciprocals)	$N(N-1)(N-2)$ $(N-3)/8\ (\alpha + \beta + \gamma)$ data	107,880	122,760
Comprehensive, effective (whole compreh.—redundant)	$(\alpha + \beta)$ data (no γ + no noisy)	68,621	80,738
Dipole-dipole β ($a = 1$, $n = 1$–6)	$((N-3) + (N-8)) \times 3$	159	165
Wenner, α (CPPC)	$(N-1)(N-2)/6$	155	165

Redundant data include configurations of γ and noisy data of very low voltage.

Noisy data are those with geometric factor larger than that of dipole-dipole configuration of maximum dipole factor of 6.

Ten different borehole surveys of standard, nonstandard, and optimized configurations are investigated here (Table 3, Figure 2). The vertical (v) circulating (c) configurations α vc and βvc represent the comprehensive data sets with all possible α and β electrode combinations, respectively. Their corresponding subconfigurationsαvcs (s = symmetrical around the midpoint) and βvcs include only the conventional symmetrical arrangements of Wenner and Schlumberger, and dipole-dipole, respectively. The configurations βl and βh represent lateral (l) and horizontal (h) bipole-bipole configurations, where their C- and P-pairs are split between the two boreholes (CP-CP). The configuration βl represents the comprehensive data set acquired in all possible electrode combinations and orientations of this bipole-bipole (CP-CP) arrangement. Its subset βh is conducted with horizontal current electrodes (flows) only.

Table 3: Applied configurations for two vertical arrays, each of 16 electrodes (see Figure2). The corresponding electrodes in each borehole are set at equal depths with unit interval

Configuration	Data no.	Characteristics
(1) tp-p	813	Vertical current flow, horizontal resolution
(2) βh	1,240	Horizontal/lateral current flow, better data quality, better vertical (less lateral) resolution
(3) βl	14,440	
(4) βhtp	2,053	Sum of βh and tp-p ((1) + (2))
(5) αvcs	1,796	Sum of Wenner and Schlumberger
(6) βvcs	1,177	Dipole-dipole
(7) αβvcs	2,973	Sum of αvcs and βvcs ((5) + (6))
(8) αvc	34,788	Comprehensive including nonstandard and standard (αvcs and βvcs in αvc and βvc, resp.) configurations
(9) βvc	31,510	
(10) opt	3,000	Optimized data set with less than 4% size but 97% resolution relative to that of comprehensive set

c: circulating, v: vertical, α, β: tripotential configuration, l: lateral, s: symmetrical, tp-p: tripole-pole, h: horizontal, opt: optimized.

The tripole-pole (tp-p) is a special type of vertical circulating crosshole configurations [24]. It has fixed C electrodes (the respective upper- and lowermost electrode of the first and second hole, and vice versa; Figure 2) and moving P electrodes between any other possible combinations of electrode pairs in each borehole separately (i.e., CPP-C, C-PPC, PPC-C, and C-CPP of Table 1). It shows near-vertical current flows that are strongly influenced by horizontal layers in the interwell region. As opposed to tp-p, bipole-bipole configurations (βl and βh) have mostly lateral current flows, that is, a low resolution for horizontal structures, but detect better vertical structures. The new complex data sets βhtp (sum of βh and tp-p) and αβvcs (sum ofαvcs and βvcs) are introduced for the first time in this study. Each should reflect the advantages of its constituting arrays. Their combined current flows and potential measurements in various possible combinations and directions should be able to resolve targets of varying orientations. The applied opt data set (3000 data points) is less than 4% of the comprehensive set but has an average relative resolution of 0.96%

(Table 3).

SUBSURFACE MODEL SCENARIOS

To keep the virtual CO_2 sequestration modeling more realistic, the formation parameters of the starting subsurface scenarios used to generate the synthetic data have been taken from published data, for example, CO_2 SINK test site of Ketzin, near Berlin (e.g., [5, 27, 28]). The applied single subsurface models consist of the electrically almost insulating, supercritical CO_2 plume ($P_{CO_2} \approx \infty$) sequestrated at the top of a conductive saline sandstone reservoir ($\rho_{reservoir} = 3\,\Omega m$, $\rho_{brine} = 0.20\,\Omega m$, salinity 35–55 g/L, $\Phi = 20$–25%), (Figure 3). This reservoir is capped by an impermeable siltstone ($\rho = 8\,\Omega m$ with varying thicknesses in the range 2a–9.5a (a = electrode spacing)) to prevent upward CO_2 leakages. Model dimensions in this study are given in unit a of which is often assumed to be 1 m. The CO_2 plume is simulated by the common wedge shape with bulk resistivities, ρ_{plume} of 100, 30, 15, and 10 Ωm (corresponding to saturations, S_{CO_2} of 80, 60, 40, and 30%, resp., as calculated from (2), and varying thicknesses (0.5a–13a) and widths (0.5a–13a)) [29].

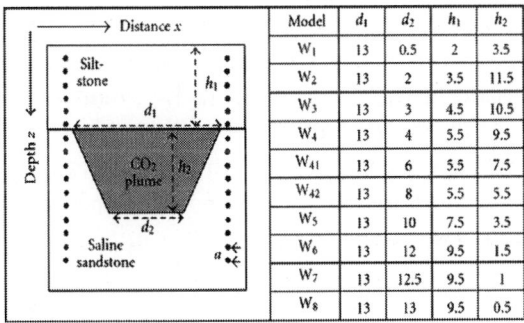

Model	d_1	d_2	h_1	h_2
W_1	13	0.5	2	3.5
W_2	13	2	3.5	11.5
W_3	13	3	4.5	10.5
W_4	13	4	5.5	9.5
W_{41}	13	6	5.5	7.5
W_{42}	13	8	5.5	5.5
W_5	13	10	7.5	3.5
W_6	13	12	9.5	1.5
W_7	13	12.5	9.5	1
W_8	13	13	9.5	0.5

Figure 3: Subsurface model scenarios of saline sandstone reservoir (resistivity, $\rho = 3\,\Omega m$) with CO_2 wedge-like plume of varying width (d), thickness (h), depth and saturation (80, 60, 40, and 30%, i.e., $\rho = 100$, 30, 15, and 10 Ωm, resp.) capped by siltstone (8 Ωm) together with borehole electrode arrays (●) used in the survey. d_1, d_2, h_1, h_2, are given in units of electrode spacing a

APPLIED PROCEDURES

The 2.5D forward and inverse ERT modeling of deep CO_2 plume was carried out using new codes based on algorithms for shallow surveys (e.g., [30]). All codes use the half-space solution with a fine mesh grid to accurately model the whole region. The Neumann and mixed boundary conditions are used for the top surface and the side/bottom boundaries, respectively. The program optimizes automatically a sufficient number of additional mesh grids far from the electrodes (5a–10a) at these side/bottom boundaries.

An extensive numerical investigation was started by generating synthetic data sets of apparent resistivity (t_a) from 2.5D forward simulation as a function of (1) ten wedge-like models (Figure 2), (2) four S_{CO_2} values, (3) ten electrode configurations (Table 3, Figure 1), (4) seven ARs, (5) two burial depths (1a and 101a) and (6) three noise levels (1%, 3%, and 5%). This forward modeling has resulted in more than 8000 synthetic data sets for these diverse model scenarios with and without the wedge-like CO_2 plume. Each synthetic data set was filtered to remove any outliers and any ρ_a values of high geometric factor which may cause a potential leakage. The filter ensures also the absence of any equivalent, reciprocal, or γ configurations from the data set. Each filtered data set has been inverted independently eight times using different setup constraint parameters. This results in more than 100,000 tomograms for all applied synthetic data sets. The diverse setup constraints applied in the inversion mainly include regularizations with the minimization methods of least squares (L_2) or robust blocky normalization (L_1), and initial models of a constant homogeneous resistivity or an approximate inverse model.

The reliability of the forward modeling using finite-element method (with the more accurate trapezoidal elements instead of triangle ones) was confirmed before starting inversions using a homogeneous medium with a constant resistivity. Compared with this constant resistivity, the resulting deviations of the single apparent resistivities (ρ_a) for each applied synthetic data set generally do not exceed 3%. These deviations are similar to the normal error level in the real data and are considered here as a noise in the synthetic data sets.

Among the 10 examined wedge-like scenarios, this study focuses mainly on models W_1, W_{42}, and W_8 with the largest (13a), intermediate

(5.5), and least (0.5a) thicknesses for the CO_2 plume (Figure 3). The electrode coverage for the cap rock is least for W_1 (only 1a), intermediate for W_{42} (5a), and best for W_8 (11a). These CO_2 storage scenarios represent examples with optimum target dimensions (W_{42}) and problems of thin layers (plume in W_8 and reservoir below the plume in W_1) and thin widths (lowermost triangle apex in W_1). Among the eight independent inversions for each data set, only the one with the best-fitting model (least root mean square error, rms-error) is considered for further studies.

In the following, any inverted tomogram with its single target anomalies (CO_2 plume, host reservoir rock, and cap rock) is evaluated relative to the starting input model according to the following criteria:

- the reconstructed geometry, shape, and position,
- the recovered resistivity magnitude,
- the sharpness of the boundaries.

RESULTS AND DISCUSSIONS

In the following, the reconstructed ERT tomograms for W_1, W_{42}, and W_8 scenarios will be described and discussed only as a function of the applied CO_2 plume scenarios (dimensions, S_{CO_2} or ρ), electrode configurations, ARs and noises.

The overall average rms-error for all inverted data sets approaches 0.9% with a nearly similar distribution between tomograms inverted by the robust L_1 norm and their corresponding models of the L_2 norm. For all studied data sets, every best-fitting tomogram (least rms-error almost of <0.5% and average iteration number of 5), among its independent eight inversions, was optimized with the L_1 norm. This low rms-error value is explained by the good convergence of the synthetic data sets toward the final solution. The average rms-error values are least for the lateral/horizontal configurations ($\beta l/\beta h$), intermediate for the opt and vertical circulating symmetrical configurations (αvcs, βvcs, $\alpha \beta vcs$), and highest for the comprehensive (αvc, βvc), tp-p and βhtp. One may note that the misfit distribution in these arrays is directly proportional to the data number with singularity problems and/or high geometric factors (i.e., of voltage leakages).

In all studied cases, the L_1 tomograms always show sharper boundaries and better magnitude recovery for the single targets than the L_2 tomograms, that is, L_1 inversion fits better for resolving sequestration targets with sharp boundaries. These results are in accordance with the evidence that the L_1 norm tends to produce models that are piecewise constant, whereas the L_2 norm tends to smear out the sharp boundaries [19].

Briefly every best resolved output tomogram (among its eight independent inversions) resulted from the set of inversion constraint parameters which incorporates the use of (1) accurate calculation options at the expense of computation time (e.g., the standard Gauss-Newton algorithm to (re-)calculate the Jacobian matrix and optimization, 4 nodes per a and mesh grid to reduce singularity errors, the model refinement of half width and crosshole model of half size, etc.), (2) a robust blocky L_1 norm for sharp boundaries instead of smooth L_2 norm for gradual changes, (3) sufficient boundary meshes (>5a), and (4) low damping parameters for these synthetic data (almost noise free). Based on these discussions, all (best-fitting) tomograms considered throughout the next sections were inverted using the robust L_1 norm.

EFFECT OF CONFIGURATIONS AND MODEL SCENARIOS

Figure 4 shows examples of the reconstructed 2D tomograms (with the best-fitting least rms-error) for the subsurface scenarios W_1, W_{42}, and W_8 of different CO_2 plume dimensions and only the two extreme S_{CO_2} of 80 and 30 g/L (corresponding ρ of 100 and 10 Ωm, resp., Figure 3). These scenarios are situated at a depth of 101a (to the uppermost electrode). They are reconstructed for the applied 10 electrode configurations (Table 3, Figure 2) using the aspect ratio (AR) of 1.

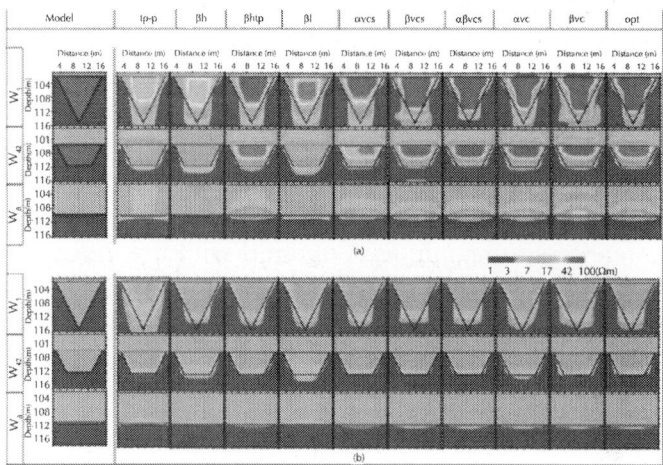

Figure 4: Numerical resistivity tomograms (3rd–12th column) inverted for different wedge-like CO_2 plumes (W_1, W_{42}, W_8, 1st-2nd column, see Figure 3) of varying saturations—resistivity of 100 (a) and $10\,\Omega m$ (b)—and 4-pole electrode configurations (1st row, see Figure 2 and Table 3 for symbols) between two borehole electrode arrays (•). All models show root mean square errors of less than 1%. Solid lines refer to target boundaries.

In most applied cases, the resistive wedge-like CO_2 plume together with the conductive saline reservoir of hosting sandstone and the impermeable siltstone cap rock are generally mapped directly by the inverted absolute ρ tomograms (no model differencing between tomograms with and without plume). Most configurations generally reconstruct these three targets with varying degrees of smearing and artifacts. This smearing is reflected on the single tomograms by target anomalies of lower magnitude, larger volume, and blurred boundary relative to that of their input models. This smearing is evaluated quantitatively by the ROI and residual analyses, see next section. The mapping capability generally decreases with decreasing dimensions (thickness, width) and resistivity of the targets. With the exception of the lateral/horizontal (βl/βh), all other configurations can reconstruct even the worst CO_2 plume scenario (of the least thickness of W_8 and resistivity of $10\,\Omega m$). Based on the previously mentioned criteria for evaluating the inverted tomogram anomalies (geometry, position, ρ amplitude, and boundary sharpness), the mapping capability generally is better for configurations with multiply orientated current flows and potential measurements (αvc and βvc, their subsets αvcs, βvcs, and $\alpha\beta$vcs, as well as opt and βhtp) than those of only lateral (βl), horizontal (βh), and vertical (tp-p) current injection. Tomographic anomalies of the later configurations generally show lower ρ magnitude than that of the former configurations, that is, configurations βl, βh and tp-p underestimate the ρ magnitude compared with the input models, see next section.

Among the applied scenarios, the mapping capability for the sequestration targets (cap rock, plume, and reservoir) is best for W_{42}, intermediate for W_1, and least for W_8. Tomogram W_1 shows strong smearing effects for the lowermost apex of plume triangle and the cap rock with their surroundings. Both targets have the thin layer problem with electrode coverage of ≤ 1 which causes this smearing governed by the equivalence principle (e.g., [15]). For a thin resistive CO_2 plume target, the inversion code is not able to resolve thickness and resistivity parameters of certain resistive model features individually but only their product (resistance). The horizontal boundaries of the upper reservoir/plume and lower plume in W_{42} and partly W_1 are generally resolved slightly better than the inclined boundaries of the plume.

Regarding the applied configurations, the vertical circulating αvc and αvcs generally resolve inclined boundaries better than their corresponding βvc and βvcs. This may explain the occasional moderate resolution of $\alpha\beta$vcs tomograms (containing βvcs data) for the lower triangle apex of the plume in W_1. It is clear that the individual complex configurations βhtp (sum of βh and tp-p) and $\alpha\beta$vcs (sum of αvcs and βvcs) combine the features (mostly of better resolution) of their corresponding single constituents. Each complex configuration contains more data measured in more orientations and thus carries more information than its constituent data sets (cf. [31]). On the other hand, models of the vertical circulating β (βvc and βvcs) tomograms generally reflect higher (better) resistivity magnitudes of the targets than their corresponding α (αvc and αvcs) models. The resistivity magnitude approaches its real value in the target centre and deviates increasingly from this with distance toward the contact with the next target. As opposed to real data, the inversion models do not show the usual poor resolution with strong artifacts around the boreholes. These are due to the heterogeneities caused by boring and electrode installation and/

or the distorting effects of the conductive borehole fluid relative to the resistive host rock [32].

Regarding the data set size, the resolution of opt tomograms (nearly 3,000 data points) is almost similar to the best possible resolution of each of the comprehensive αvc and βvc models (each of >30,000 points) and far better than that of βl tomograms (14,440). Obviously, the resolution for the noncomprehensive data sets is best for opt, above moderate for the complex αβvcs and βhtp, moderate for αvcs and βvcs, and least for βh and tp-p.

Briefly, the lateral/horizontal configurations (βl and βh) are more robust against measurement errors due to current and voltage leakages. In these configurations, splitting the current and potential electrode pair between boreholes (i.e., CP-CP) maximizes the measured voltage (e.g., [23]). Large borehole offset (i.e., P_1-P_2 distance), however, may include the telluric noise in the data (e.g., [33]). This noise can be minimized by periodically reversing the current flow in the current electrodes. These arrays result in poorly resolved anomalies, especially for lateral boundaries, often with underestimated ρ magnitudes. This is due to the predominance of current flows and potential measurements in the lateral directions with poor data coverage in the vertical one. Configurations having either a C or P electrode pair or both in the same borehole (inhole) such as most of the vertical circulating α and β (αvc, βvc, αvcs, and βvcs, Table 1) configurations have the singularity problem of very low voltages. However, their filtered data sets after removing this noise can resolve well the subsurface targets of CO$_2$ plume, reservoir, and cap rock assuming enough coverage of more than one electrode spacing. These conclusions are in accordance with that of, for example, Oldenburg and Li [16] and Oldenborger et al. [34] and opposed to that of Zhou and Greenhalgh [23]. In solute transport experiments, the former authors found that the vertical βvcs tomograms show more reliable subsurface structures than the horizontal βtomograms. Bing and Greenhalgh stated that the acquisition data for vertical inhole configurations are easily obscured by background noise and yield images inferior to those from lateral/horizontal configurations βl andβh.

EVALUATION OF RECONSTRUCTED TOMOGRAMS

This evaluation is carried out by the ROI analysis and the model difference (residual) relative to the input model. The ROI analysis is started by conducting two independent inversions of the same dataset using different resistivity values (q_A and q_b) for homogeneous reference models. Values of q_A and q_b are calculated from the average logarithmic ρ_a using two different multiplication factors. The inversion results from these two starting models are used to calculate the ROI value for each pixel, defined as [35]:

$$\text{ROI}(x, z) = \left[q_A(x, z) - q_B(x, z) \right] \left(q_A - q_B \right)^{-1}. \tag{3}$$

The ROI approaches zero when the model is well constrained by the data (the two inversions reproduce very similar ρ values) and one when the model has a very poor data coverage. The two inversions of each ROI analysis were conducted using the same set of inversion setup parameters (resulted in the best fitting tomograms, see before) and only three iterations. Higher iteration numbers were tested and resulted in artifacts as the algorithm tries to reduce the data misfit by modeling the noise as well. Different pairs of multiplication factors (0.1 and 10, 0.1 and 10, and 0.2 and 5) were applied here and yield similar results (see also [34, 36]).

On the other hand, the relative resistivity difference or residual ($\Delta\rho$) between each corresponding pixel of the input (ρ_{input}) and output (ρ_{output}) 2D models is calculated by

$$\Delta\rho = \left(\rho_{output} - \rho_{input} \right) \left(\rho_{output} + \rho_{input} \right)^{-1} \tag{4}$$

Moreover, a random noise at 1, 2, and 5% levels was added to the synthetic data sets, in addition to their forward modeling errors (up to 3%), to evaluate noisy field effects on the inversion. Adding noise in this order generally increases the rms-error values by a factor of 2 to 9 but slightly decreases the mapping capability, particularly for W_1 and W_{42}. Ramirez et al. [20] obtained similar results; the effect of the random error is insignificant for anomalies of a large size and magnitude.

Figure 5 shows examples of the resulting ROI and Δρ tomograms as a function of the applied electrode configurations carried out for the subsurface scenarios W$_1$, W$_{42}$, and W$_8$ at 101a depth only. Constraining of inverted tomograms by data coverage is best (lowest values for ROI and partly Δρ) for the comprehensive αvc and βvc, and opt arrays, intermediate for the complex αβvcs and βhtp, and poor (highest ROI values) for the others. This confirms the effectiveness of our optimization approach applied here to generate a practical opt dataset of high resolution.

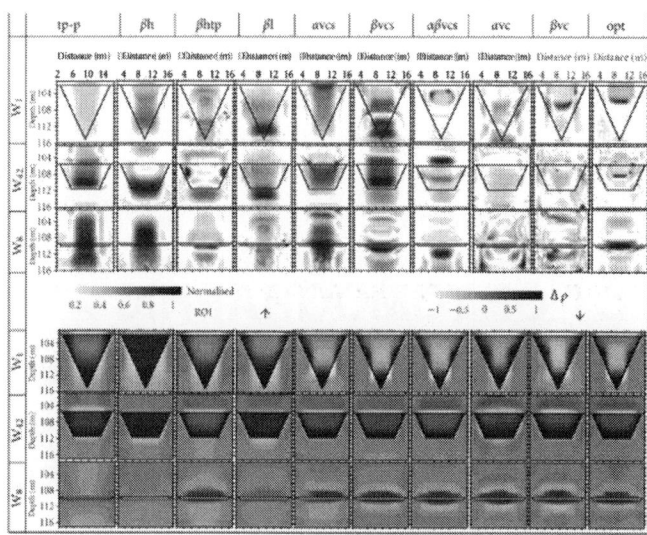

Figure 5: Evaluation of the reconstructed electrical resistivity tomograms (depth of 101a, = electrode spacing) using methods of region of index, ROI (top, (3)) and relative model difference, Δρ (bottom, (4)) for survey between two borehole electrode arrays (•) as a function of subsurface scenario (W$_1$, W$_{24}$, and W$_8$, 1st column), and electrode configuration (see Table 3 for symbols). Solid lines refer to target boundaries.

Unlike real field data, this synthetic modeling analysis does not show the usual disadvantageous low resolution of high values of ROI and Δρ (of bad data and thus resolution) around the boreholes due to the heterogeneities resulting from boring and electrode installation. The highest values of ROI and Δρ with the least data coverage and resolution are concentrated in the central interwell region which is a common disadvantage for ERT results. Also ROI and Δρ analyses reflect similar results between the corresponding single tomograms of shallow (1a depth, not shown here) and deep scenarios (101a depth). This similarity reflects the reliability of the applied techniques. A comparison with published results (e.g., [34]) shows that the vertical configurations reflect better resolution (of lower ROI values) than the lateral ones which confirms the results obtained here.

Briefly, tomograms of Δρ and ROI generally show similar results regarding the mapping capability of the single applied arrays. They reflect well the common smearing effects of varying degrees. These effects lead to overpredicted volumes, underpredicted magnitudes, and blur boundaries of the target anomalies. This study shows clearly that the complex (βhtp and αβvcs) arrays with multiply oriented measurements in addition to opt arrays with practical data sizes are recommended for highly spatiotemporal resolution and will be considered further in this study.

EFFECT OF TOMOGRAPHIC ASPECT RATIO (AR)

Extensive tests for seven different AR steps within 0.25–2.0 range were performed as a function of the applied subsurface scenarios (Figure 2), S_{CO_2}, burial depths, electrode configurations (Figure 1, Table 3), noises, and modeling setup parameters. At the AR values (0.5, 0.7, 1.0, 1.3 and 2), Figure 6 shows some best-fitting (least rms-error) tomograms resulting from unconstrained inversions only for the highly resolving arrays (βhtp, αβvcs and opt) and the subsurface scenario W_{42} with S_{CO_2} of 60% ($\rho_{co2}=30\Omega m$).

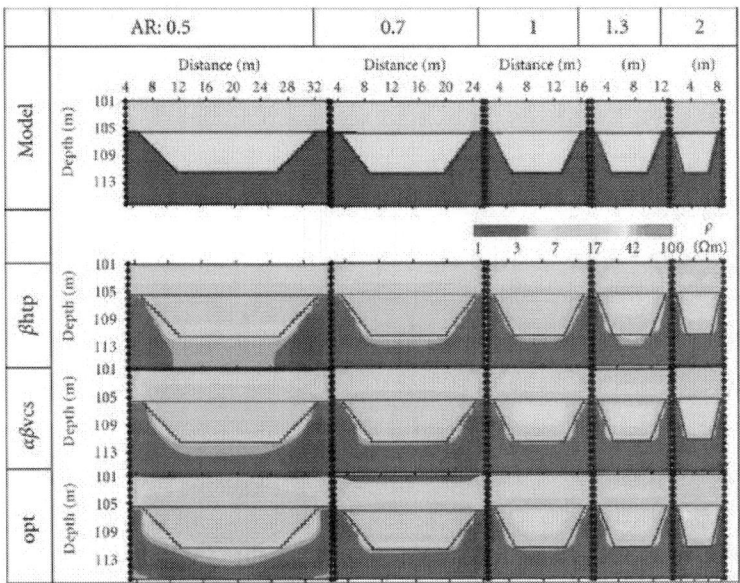

Figure 6: Effect of aspect ratios, AR (first row) on ERT tomograms, resulting from wedge-like CO$_2$ plume scenario W$_{42}$ (Figure 3, second row) with 30 Ωm plume resistivity for 4-pole configurations βhtp, αβvcs and opt (see Figure 2 and Table 3 for symbols) between two borehole electrode arrays (●). All models show root mean square errors of less than 1%. Solid lines refer to target boundaries.

Figure 6 shows that the obtained mapping capability (including resolution) generally increases with increasing AR from 0.5 to 2. This occurs although the lateral boundaries of the CO$_2$ wedge plume become more vertical with increasing AR. This mapping improvement with increasing AR is also associated with a slight increase of the recovered magnitude of the CO$_2$ plume relative to that of the starting input model. Most other results of the previous sections are manifested here such as the poor mapping resolution of thin layers (of W$_1$ and W$_8$scenarios, not shown here). At AR = 1, the lowermost triangle apex of W$_1$ is predominated either from the reservoir anomaly or from a wide smeared zone of the plume. At AR>1, the apex resolution increases with increasing AR and approaches best results at AR of 2. Based on the applied electrode configurations, the resolution for tomograms of the vertical circulating and opt arrays is better than for those of lateral/

horizontal$\beta l/\beta h$ and partly tp-p and βhtp configurations (partly not shown here). Compared with W_1 and W_8, W_{42} models are generally better resolved due to the better coverage of their target. The ability to detect and often map the three sequestration targets (CO_2 plume, reservoir, and cap rock) by unconstrained inversions is still possible with AR values down to 0.5 for the most studied scenarios (even those with the worst scenario of least thickness and ρ). This result is superior to that of published studies (e.g., [18]). These authors applied ERT techniques for site characterization and process monitoring and determined a minimum AR with acceptable output resolution of 1. In comparison, the minimum AR value (0.5) currently obtained in this study can lead to a decrease of the number of the expensive monitoring wells and the costs by a factor of 4 (i.e., $n^{(1/AR)}$, n:well number for AR = 1). The reconstructed output tomograms for lower AR values (<0.5) achieve a satisfactory resolution only for constrained inversions with an a priori fixing of boundaries and/or resistivities of the targets. The resolution increases with increasing the number of constraints.

SUMMARY AND CONCLUSIONS

Electrical resistivity tomography (ERT) techniques in boreholes are powerful in monitoring intrinsic property changes for storing the resistive (supercritical) CO_2 in conductive saline reservoirs. In this study, the mapping capability of various ERT techniques is studied for diverse wedge-like CO_2 plumes in a deep saline aquifer capped by an impermeable rock. Extensive, systematic 2.5D modeling studies (>100,000 models) were performed to test the ERT sensitivity for a multitude of parameters related to the subsurface setting (hydrogeology and geochemistry of reservoir and cap rock), CO_2 plume reservoir, survey design, data acquisition, and modeling techniques. The new array optimization approach is applied to generate optimized data sets (opt) of only 4% of the comprehensive set but of almost similar resolution. Forward simulations were carried out to generate diverse synthetic data sets (>8000) as a function of plume scenarios (different dimensions and CO_2 saturations S_{CO_2} or resistivity, ρ), burial depths, electrode configurations, random noise, and aspect ratios (AR). The data quality (<3% noises) is confirmed by results of tests on a homogeneous model with constant ρ. This numerical study principally reveals the capability of ERT techniques to resolve the various deep subsurface scenarios with the CO_2 sequestration targets (plume, host reservoir, and cap rock). Most important results may be summarized as follows:

- Most applied ERT configurations can generally map the sequestration targets of sharp boundaries directly by the absolute ρ tomograms (no model differencing) using L1 robust inversions. All models, however, reflect smeared anomalies of lower magnitude, larger area, and blurred boundary.

- Superior to published studies, the detection of CO_2 targets is possible even for the worst scenario of 0.5a thicknesses (a = unit electrode length), 30% S_{CO_2}, and 0.5 AR. At lower AR values (<0.5), a satisfactory resolution can result only from constrained inversions with an a priori fixing of boundaries and/or resistivities of targets.

- The developed opt and complex (αβvcs, βhtp) arrays (nearly 3000 data points) are recommended for surveys of highly spatiotemporal resolution. Their resolution is the second best after the comprehensive vertical arrays (αvc and βvc, each of >30,000 data) and far better than the comprehensive lateral array (βl, 14,400 data). These arrays improve the common low resolution of the ERT technique in the interwell region by combining configurations with current flows and potential measurements in all possible orientations.

- Lateral arrays (βl and βh) are more robust against measurement noise; their synthetic data sets result in tomograms with the least rms-error misfit, but their real field data may include natural telluric noise at large borehole offset.

- Analyses of the region of index, ROI, and residual model difference relative to the input model show that inverted tomograms of comprehensive (αvc and βvc), opt, and complex (αβvcs) configurations are better constrained by the data than those of the other applied ones. This tomogram reliability generally increases by increasing the data size.

- Sometimes the configuration tripole-pole (tp-p) is able to detect horizontal structures due to its near-vertical current flows. Contrary to this, lateral arrays (βl and βh) have lateral current flows and thus better resolution for vertical structures.

- The vertical circulating α and β data sets are collected partly with either or both of the current and potential electrode pair inhole and thus may show singularity problems. Filtering these data sets results in tomograms of better resolution than those of lateral arrays.

- Adding random noise (1%, 3%, and 5%) to the synthetic data (in addition to its forward modeling errors, up to 3%) increases the rms-error values (according to the error continuation law) by a factor of 2–9 but slightly decreases the mapping capability of the techniques, particularly for large targets.

OUTLOOK AND RECOMMENDATION

The current results give answers to some studied problems of 2D ERT mapping for CO_2 sequestration in deep saline aquifers. Many other problems related to the 2D/3D mapping and 4D monitoring are currently studied within the research activities of our MoPa project. These tasks include the following:

- Continuing the systematic 2.5D modeling studies for other CO_2 plume scenarios and extending these to 3D investigations with varying parameters related to modeling, data acquisition and methodology, geological setting and plume reservoir developments based on petrophysical approaches.

- Applying approaches of time-lapse imaging for 4D monitoring of any CO_2 migration either laterally and downwards within the saline reservoir or upwards through probable postinjection fracturing of the cap rock caused by the high injection pressure. This includes monitoring of pre- and postinjection scenarios, developments of the new CO_2 reservoir, and any change in the porosity/permeability with time.

- Risk analyses for any possible postinjection fracturing in the cap rock and upward CO_2 seepages into the near surface zone and especially freshwater aquifers.

- Evaluating the spatiotemporal resolution of output tomograms using techniques of residual, 2D ROI, and 3D VOI. Quantitative analyses of ρ magnitude and spatial extension in the single pixels/voxels of the recovered tomograms with respect to the pre-injection model.

- Applying developed modeling techniques on more realistic subsurface scenarios by using available (meta)data of the North German Basin and ρ_a (hard data) inversion constrained by seismic and log information (soft data) to reduce the problems of nonuniqueness. Constrained inversion schemes should be able to fix resistivity regions and include boundaries in the output models.

- Using petrophysical approaches to quantify inverted resistivity models in saturation, and the quality of the pore filler (brine and CO_2).

- Refining and verifying the use of optimized configurations for enhancing the spatial resolution without a significant decline of temporal resolution.

- Testing the developed techniques on real field data, for example, the sequestration site of Ketzin in collaboration with partners of the project CO_2 sink.

ACKNOWLEDGMENTS

Special thanks to colleagues M. H. Loke for providing modeling programs, M. Strahser, W. Rabbel, and R. Meissner for critical comments, T. Wunderlich, S. Siebrands, and A. Ismaeil for MATLAB programs, and Associate Editor X. Yang and two anonymous reviewers for constructive recommendations. This study is funded by the German Federal Ministry of Education and Research (BMBF), EnBW Energie Baden-Württemberg AG, E.ON Energie AG, E.ON Gas Storage AG, RWE Dea AG, Vattenfall Europe Technology Research GmbH, Wintershall Holding AG, and Stadtwerke Kiel AG as part of the CO_2-MoPa joint project in the framework of the Special Programme GEOTECHNOLOGIEN.

REFERENCES

1. M. Meadows, "Time-lapse seismic modeling and inversion of CO_2 saturation for storage and enhanced oil recovery," Leading Edge, vol. 27, no. 4, pp. 506–516, 2008. ··

2. S. A. Hagrey, "First numerical ERT models for CO_2 plumes in saline reservoirs using crosshole configurations," in Proceedings

of the 1st EAGE CO2 Geological Storage Workshop, p. 5, European Association of Geoscientists and Engineers, 2008.

3. G. E. Archie, "The electrical resistivity log as an aid in determining some reservoir characteristics,"Transactions of the American Institute of Mining Engineers, vol. 146, pp. 54–62, 1942.

4. B. A. Kirkendall and J. J. Roberts, "Crosswell electromagnetic imaging, in advanced reservoir characterization in the Antelope shale to establish the viability of CO_2 enhanced oil recovery in California's Monterey formation siliceous shale," Annual Report DOE/BC/14938-12, National Petroleum Technology Office, 2003.

5. A. Förster, B. Norden, K. Zinck-Jørgensen et al., "Baseline characterization of the CO_2SINK geological storage site at Ketzin, Germany," Environmental Geosciences, vol. 13, no. 3, pp. 145–161, 2006. · ·

6. A. L. Ramirez, J. J. Nitao, W. G. Hanley et al., "Stochastic inversion of electrical resistivity changes using a Markov Chain Monte Carlo approach," Journal of Geophysical Research B: Solid Earth, vol. 110, no. 2, pp. 1–18, 2005. · ·

7. J. Van Sijl, P. Winthaegen, B. Goes, and C. J. Peach, "Modelling ERT monitoring of CO_2-storage with enhanced coalbed methane recovery," in Proceedings of the 69th European Association of Geoscientists and Engineers Conference and Exhibition, pp. 1220–1224, London, UK, June 2007.

8. F. D. Day-Lewis, K. Singha, and A. M. Binley, "Applying petrophysical models to radar travel time and electrical resistivity tomograms: resolution-dependent limitations," Journal of Geophysical Research B: Solid Earth, vol. 110, no. 8, Article ID B08206, pp. 1–17, 2005. · ·

9. P. Stummer, H. Maurer, and A. G. Green, "Experimental design: electrical resistivity data sets that provide optimum subsurface information," Geophysics, vol. 69, no. 1, pp. 120–139, 2004.

10. T. Hennig, A. Weller, and M. Möller, "Object orientated focussing of geoelectrical multielectrode measurements," Journal of Applied Geophysics, vol. 65, no. 2, pp. 57–64, 2008. · ·

11. P. B. Wilkinson, P. I. Meldrum, J. E. Chambers, O. Kuras, and R. D. Ogilvy, "Improved strategies for the automatic selection of

optimized sets of electrical resistivity tomography measurement configurations,"Geophysical Journal International, vol. 167, no. 3, pp. 1119–1126, 2006. · ·

12. P. R. McGillivray and D. W. Oldenburg, "Methods for calculating Frechet derivatives and sensitivities for the non-linear inverse problem: a comparative study," Geophysical Prospecting, vol. 38, no. 5, pp. 499–524, 1990.

13. I. Coscia, L. Marescot, H. Maurer, S. Greenhalgh, and A. G. Green, Experimental Design for Cross Hole Electrical Resistivity Tomography Datasets, EAGE Near Surface Geophysics, Cracow, Poland, 2008.

14. S. A. Hagrey, 2D Optimisation of Electrode Arrays for Borehole Surveys, EAGE Near Surface Geophysics, Dublin, Ireland, 2009.

15. S. A. Hagrey, "2D optimized electrode arrays for borehole resistivity tomography and CO_2 sequestration modelling," Pure and Applied Geophysics. In press. ·

16. D. W. Oldenburg and Y. Li, "Estimating depth of investigation in dc resistivity and IP surveys,"Geophysics, vol. 64, no. 2, pp. 403–416, 1999.

17. M. S. Zhdanov, Geophysical Inverse Theory and Regularization Problems, Elsevier, New York, NY, USA, 2002.

18. R. L. Newmark, W. Daily, and A. Ramirez, "Electrical resistance tomography using steel cased boreholes as electrodes," in Proceedings of the 69th Annual Meeting of Expanded Abstract, p. 4, Society of Exploration Geophysicists, 1999.

19. N. B. Christensen, D. Sherlock, and K. Dodds, "Monitoring CO_2 injection with cross-hole electrical resitivity tomography," Exploration Geophysics, vol. 37, no. 1, pp. 44–49, 2006.

20. A. L. Ramirez, R. L. Newmark, and W. D. Daily, "Monitoring carbon dioxide floods using Electrical Resistance Tomography (ERT): sensitivity studies," Journal of Environmental and Engineering Geophysics, vol. 8, no. 3, pp. 187–208, 2003.

21. S. A. Al Hagrey and T. Petersen, "Numerical and experimental mapping of small root zones using optimized surface and borehole resistivity tomography," Geophysics, vol. 76, no. 2, pp. G25–G35, 2011. ·

22. E. W. Carpenter and G. M. Habberjam, "A tripotential method for resistivity prospecting," Geophysics, vol. 21, pp. 455–469, 1956.

23. B. Zhou and S. A. Greenhalgh, "Cross-hole resistivity tomography using different electrode configurations," Geophysical Prospecting, vol. 48, no. 5, pp. 887–912, 2000. · ·

24. B. J. M. Goes and J. A. C. Meekes, "An effective electrode configuration for the detection of DNAPLs with electrical resistivity tomography," Journal of Environmental and Engineering Geophysics, vol. 9, no. 3, pp. 127–141, 2004.

25. M. Noel and B. Xu, "Archaeological investigation by electrical resistivity tomography: a preliminary study," Geophysical Journal International, vol. 107, no. 1, pp. 95–102, 1991.

26. M. H. Loke, F. A. Fouzan, and M. N. M. Nawawi, "Optimisation of electrode arrays used in 2D resistivity imaging surveys," in Proceedings of the 19th Conference and Exhibition, p. 5, Australian Society of Exploration Geophysicists, 2007.

27. A. Ramirez, J. Friedmann, W. Foxall, et al., "oint reconstructions of CO_2 plumes using a Markov Chain Monte Carlo approach," in Proceedings of the 8th International Conference on Greenhouse Gas Control Technologies, p. 6, Tondheim, Norway, 2006.

28. D. Kiessling, C. Schmidt-Hattenberger, H. Schuett et al., "Geoelectrical methods for monitoring geological CO_2 storage: first results from cross-hole and surface-downhole measurements from the CO_2SINK test site at Ketzin (Germany)," International Journal of Greenhouse Gas Control, vol. 4, no. 5, pp. 816–826, 2010.

29. E. Gasperikova and G. M. Hoversten, "Gravity monitoring of CO_2 movement during sequestration: model studies," Geophysics, vol. 73, no. 6, pp. WA105–WA112, 2008. ·

30. M. H. Loke, I. Acworth, and T. Dahlin, "A comparison of smooth and blocky inversion methods in 2D electrical imaging surveys," Exploration Geophysics, vol. 34, pp. 182–187, 2003.

31. A. Furman, T. P. A. Ferré, and G. L. Heath, "Spatial focusing of electrical resistivity surveys considering geologic and hydrologic layering," Geophysics, vol. 72, no. 2, pp. F65–F73, 2007. · ·

32. J. A. Doetsch, I. Coscia, S. Greenhalgh, N. Linde, A. Green, and T. Günther, "The borehole-fluid effect in electrical resistivity imaging," Geophysics, vol. 75, no. 4, pp. F107–F114, 2010. · ·

33. M. Blome, H. Maurer, and S. Greenhalgh, "Geoelectric experimental design—efficient acquisition and exploitation of complete pole-bipole data sets," Geophysics, vol. 76, no. 1, pp. F15–F26, 2011. ·

34. G. A. Oldenborger, P. S. Routh, and M. D. Knoll, "Model reliability for 3D electrical resistivity tomography: application of the volume of investigation index to a time-lapse monitoring experiment,"Geophysics, vol. 72, no. 4, pp. F167–F175, 2007. · ·

35. L. Marescot and M. H. Loke, "Using the depth of investigation index method in 2D resistivity imaging for civil engineering surveys," in Proceedings of the Symposium on the Application of Geophysics to Engineering and Environmental Problems (SAGEEP ‹04), pp. F589–F595, Colorado Springs, Colo, USA, February 2004.

36. J. E. Nyquist, J. S. Peake, and M. J. S. Roth, "Comparison of an optimized resistivity array with dipole-dipole soundings in karst terrain," Geophysics, vol. 72, no. 4, pp. F139–F144, 2007. · ·

Spillways Scheduling for Flood Control of Three Gorges Reservoir Using Mixed Integer Linear Programming Model

Maoyuan Feng[1,2] and Pan Liu[1,2]

[1]State Key Laboratory of Water Resources and Hydropower Engineering Science, Wuhan University, Wuhan 430072, China

[2]Hubei Provincial Collaborative Innovation Center for Water Resources Security, Wuhan 430072, China

ABSTRACT

This study proposes a mixed integer linear programming (MILP) model to optimize the spillways scheduling for reservoir flood control. Unlike the conventional reservoir operation model, the proposed MILP model specifies the spillways status (including the number of spillways to be open and the degree of the spillway opened) instead of reservoir release, since the release is actually controlled by using the spillway. The piecewise linear approximation is used to formulate the relationship between the reservoir storage and water release for a spillway, which should be open/closed with a status depicted by a binary variable. The control order and symmetry rules of spillways are described and incorporated into the constraints for meeting the practical demand. Thus, a MILP model is set up to minimize the maximum reservoir storage. The General Algebraic Modeling System (GAMS) and IBM ILOG CPLEX Optimization Studio (CPLEX) software are used to find the optimal solution for the proposed MILP model. The China's Three Gorges Reservoir, whose spillways are of five types with the total number of 80, is selected as the case study. It is shown that the proposed model decreases the flood risk compared with the conventional operation and makes the operation more practical by specifying the spillways status directly.

INTRODUCTION

Flood disasters, accounting for about one-third of all natural catastrophes throughout the world, have been extremely severe in recent decades [1]. For example, flood disasters have caused the loss of 30 billion dollars per year in China [1–3]. As a result, reservoirs have been built and served for one of the most useful measurements for flood control.

Reservoir operations are complex, nonlinear control processes and significantly affected by hydrological conditions and constraints, which are not predictable beforehand [4, 5]. Great effort has been made to determine the optimal scheduling of the reservoirs with various methods and techniques, including linear programming, nonlinear programming, dynamic programming, and genetic algorithm [3–28]. Karaboga et al. [6] proposed a control method to derive reservoir operating rules based on the fuzzy logic with optimum rule number and tabu search. Wei and Hsu [7] presented the tree-based rules which were used to determine the optimal real-time releases for a multipurpose multireservoir system during flood periods. Bagis and Karaboga [8] developed an evolutionary algorithm-based fuzzy proportional derivative-type controller for reservoir operation. Chang [9] proposed a penalty-type genetic algorithm to find a rational reservoir release hydrograph for flood control. Li et al. [10] developed a dynamic control operation model that considers inflow uncertainty. Fu [11] presented a fuzzy optimization method based on the concept of ideal and anti-ideal solutions. Hashemi et al. [12] presented a multiple attribute group decision-making model based on the compromise ratio method. Karbowski et al. [13] presented a hybrid analytic/rule-based approach to reservoir system management during flood seasons. Liu et al. [28] proposed three methods to derive the multiple near-optimal solutions to deterministic reservoir operation problems.

Based on the above methods and techniques, the reservoir water release hydrograph can be obtained. However, the reservoir operation is a control process that essentially manages the spillway gates of dams to increase or decrease the released water [29]. In practice, two basic issues associated with spillway gates should be determined: (1) the number of various spillways to be open or used and (2) the degree of the spillway opened (full or scale open). Most of the solutions proposed so far address the release scheduling problem leaving the allocation problem as a secondary one, performed by trial and error methods. This study deals with the spillways scheduling, instead of release scheduling, for the flood control reservoir, which has seldom been addressed in the literature. The most popular reservoir operation method, dynamic programming, becomes difficulty for this specified issue, owing to the large number of discrete states (say reservoir storage) and heavy computation for accuracy.

The mixed integer linear programming (MILP) model ensures a global optimal solution, which hence is widely used in optimization fields [14–18, 30, 31]. For example, Needham et al. [14] presented a MILP model for a reservoir system analysis of three projects on the Iowa and Des Moines rivers. Norouzi et al. [15] proposed a MILP model for short term unit commitment for hydro and thermal generation units with security-constrained commitment. Liu et al. [17] used a MILP model for the optimal load distribution, which reaches the global optimum, to validate the proposed algorithm in a hydropower station. Ashouri et al. [30] developed a MILP model to obtain the optimal design and operation of building services. Luathep et al. [31] proposed a MILP model for solving a mixed transportation network design problem.

This study aims at developing a MILP model to operate reservoir by scheduling spillways. In Section 2, the MILP model is set up with (1) transforming the objective function into a linear form and (2) formulating the constraints of potential maximum water release as a piecewise linear function. Section 3 describes a case study application to China's Three Gorges Reservoir (TGR), where the optimal scheduling is compared with the conventional scheduling method. Finally, conclusions are given in Section 4.

MATHEMATICAL MODEL

Reservoir Flood Control Model

The commonly used reservoir flood control model is as follows (e.g., [4, 5]).

Objective Function

For the reservoir flood control operation, maximum water storage should be minimized, that is,

$$\min\ \max\left(V_1, V_2, \ldots, V_t, \ldots, V_T\right),$$
(1)

where V_t is the reservoir water storage at time t and T is the number of time periods.

Constraints

- Reservoir water balance equation:

$$= V_t + \left(\frac{I_t + I_{t+1}}{2} - \frac{O_t + O_{t+1}}{2}\right)\Delta t, \quad t = 1, 2, \ldots,$$
(2)

where I_t and O_t are the reservoir inflow and release at time t, respectively. Δt is the time step length. It should be noted that the water losses from the reservoir in the form of seepage and evaporation are omitted in this study.

- Water storage capacity constraint:

$$\underline{V} \le V_t \le \overline{V}, \quad t = 1, 2, \ldots, T,$$

$$(3)$$

where $\underline{V}$ and $\overline{V}$ denote the minimum and maximum reservoir storages, respectively.

- Reservoir potential maximum water release constraint:

$$\mathcal{D}_t \le f(V_t), \quad t = 1, 2, \ldots, T,$$

$$(4)$$

where $f(.)$ is the functional relationship between the reservoir storage and potential maximum water release.

- Water release constraint for the downstream safety:

$$O_t \le O_{max}^{down}, \quad t = 1, 2, \ldots, T,$$

$$(5)$$

where O_{max}^{down}, often a constant, is the water release for the downstream safety.

Linearization

The objective function and all the constraints should be in a linear form for a MILP model. However, the objective function (1) and potential maximum water release constraint (4) are unsatisfied with this assumption. Consequently, transformations have been proposed as follows.

Objective Function

A new variable V_m is introduced to represent the maximum value of V_t; that is, $V_m = \max(V_1, V_2, \ldots, V_t, \ldots, V_T)$. Then the objective function can be transferred as follows:

$$\min V_m$$

$$(6)$$

with an additional constraint:

$$V_m \geq V_t, \quad t = 1, 2, \ldots, T.$$

$$(7)$$

Potential Maximum Water Release Constraint

Recalling (4), the potential maximum water release depends on the functional relationship $f(\cdot)$ and the current water storage, while the relationship is determined based on all spillways (including turbines). Since the reservoir release is the sum of all spillways, we have

$$\sum_{i=1}^{n} q_{i,t} = \sum_{i=1}^{n} g_i(V_t, S_i), \quad t = 1, 2$$

$$(8)$$

where $q_{i,t}$, namely, $g_i(V_t, S_i)$, is the release for spillway i, which can be described with the reservoir storage V_t and status S_i (closed, full open, or scale open). n is the number of spillways. It should be noted that the spillway of scale open is always limited to several specific degrees, which are denoted as

$$S_{i,t}^1, S_{i,t}^2, \ldots, S_{i,t}^{m_i},$$

where m_i is the number of possible statuses for spillway i.

- Piecewise Linear Approximation of Relationship between Water Release and Reservoir Storage for Individual Spillway. A nonlinear function can be linearized with additional binary variables [14–17], which is very common for the interpolation of the relationship between reservoir storage and water release. As shown in Figure 1, for a specific spillway i with the status $S_{i,t}^j$, the water release $q_{i,t}^j$ is a function of the reservoir storage , and this relationship is often nonlinear. Assuming that the nonlinear function is approximated with breakpoints (Figure 1), the water release can be expressed as a piecewise linear function as follows:

$$\leq w_{i,t,p}^j \leq 1, \quad p = 1, 2, \ldots, k_i^j$$

$$(9)$$

$$\sum_{p=1}^{k_i^j} w_{i,t,p}^j = 1,$$

(10)

$$\sum_{p=1}^{k_i^j-1} r_{i,t,p}^j = 1,$$

(11)

$$\leq \begin{cases} {}^{,}i,t,p & \qquad \\ r_{i,t,p}^j + r_{i,t,p-1}^j & \left(1 < p \right. \\ {}_{...}^j & \left({}_{...} - 1 \right. \end{cases}$$

(12)

$$V_t = \sum_{p=1}^{k_i^j} \left(w_{i,t,p}^j V_{i,p}^j \right),$$

(13)

$$q_{i,t}^j = \sum_{p=1}^{k_i^j} \left(w_{i,t,p}^j Q_{i,p}^j \right),$$

(14)

where $w_{i,t,p}^j$ is the weight of breakpoint p for the spillway i with status $S_{i,t}^j$ at time t and $r_{i,t,p}^j$ is the binary variable to ensure atmost two adjacent breakpoints are greater than zero. $V_{i,}^j$ and $Q_{i,}^j$ are the reservoir storage and water release for breakpoint p of spillway i with status $S_{i,t}^j$.

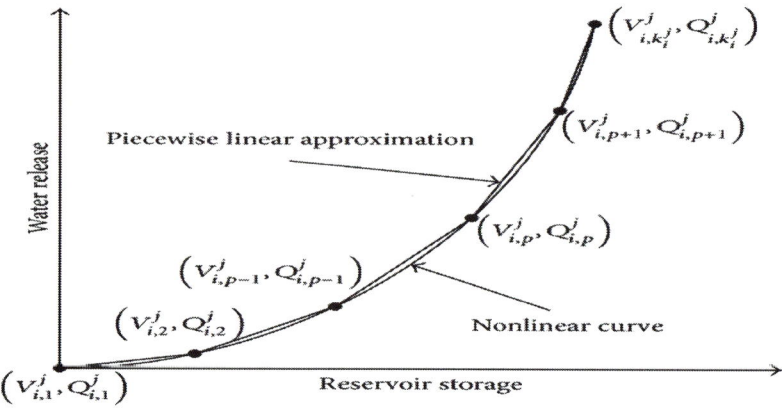

Figure 1: Piecewise linear approximation of relationship between water release and reservoir.

Equation (11) implies that only one of the binary variables $r^j_{i,t,p}$ is equal to one, and (12) ensures that two adjacent $w^j_{i,t,p}$ can be nonzero, which makes a linear interpolation between these two breakpoints. Equations (13) and (14) are the linear combinations of the reservoir storage and water release, respectively. Therefore, (10) to (14) transfer the nonlinear relationship between water release and reservoir storage into a piecewise linear function.

- Water Release for Individual Spillway. Based upon the above piecewise linear relationship, the water release of each individual spillway can be expressed as follows:

$$q_{i,t} \geq q^j_{i,t} - \left(1 - s^j_{i,t}\right) O_{\max},$$

(15)

$$q_{i,t} \leq q^j_{i,t} + \left(1 - s^j_{i,t}\right) O_{\max},$$

(16)

$$q_{i,t} \geq -s_{i,t}^{j} O_{max},$$

$$(17)$$

$$q_{i,t} \leq s_{i,t}^{j} O_{max},$$

$$(18)$$

$$\sum_{j=1}^{m_i} s_{i,t}^{j} \leq 1,$$

$$(19)$$

where $s_{i,t}^{j}$ is the binary variable to describe the status of spillway i at time t; that is, zero means that the spillway is not used with status $s_{i,t}^{j}$; otherwise this spillway opens with status $S_{i,t}^{j}$. Equations (15) to (18) form an if-then statement; that is, $s_{i,t}^{j} = 1$ means $q_{i,} = q_{i,t}^{j}$ and $s_{i,t}^{j} = 0$ means $q_{i,t} = 0$. Equation (19) ensures that only one status could be used, including the zero for closed status. Therefore, the binary variables $s_{i,t}^{j}$, can be used to indicate the status of the spillway.

- Control Order of Spillway. Two common control rules for the spillway are as follows.
- Symmetry rules: it is very popular for the spillways to open/close symmetrically, which ensures the safety of dam. For example, the spillways i_1 and i_2 should be open/closed at the same time, and this rule can be described as follows:

$$\sum_{j=1}^{m_{i_2}} s_{i_2,t}^{j} = \sum_{j=1}^{m_{i_1}} s_{i_1,t}^{j}.$$

$$(20)$$

- Control order: if the spillways i_1 should be used prior to both of i_2 and i_3; this rule can be described as follows:

$$\sum_{j=1}^{m_{i_2}} s_{i_2,t}^{j} + \sum_{j=1}^{m_{i_3}} s_{i_3,t}^{j} \leq M \sum_{j=1}^{m_{i_1}} s_{i_1,t}^{j},$$

(21)

where M is a large positive value.

Equation (4) is then reformulated with linear equations from (8) to (21) for the consideration of the spillway rules.

CASE STUDY

Three Gorges Reservoir

The Three Gorges Reservoir (TGR) is a vital project for water resources development of China's largest river, the Yangtze River (Figure 2). The TGR receives inflow from a 4.5×10^3 km long channel with a contributing drainage area of 10^6 km^2. The mean annual runoff at the dam site is 451 billion m^3. With a flood storage capacity of 22.15 billion m^3, the TGR plays the most important role in flood control of the Yangtze River.

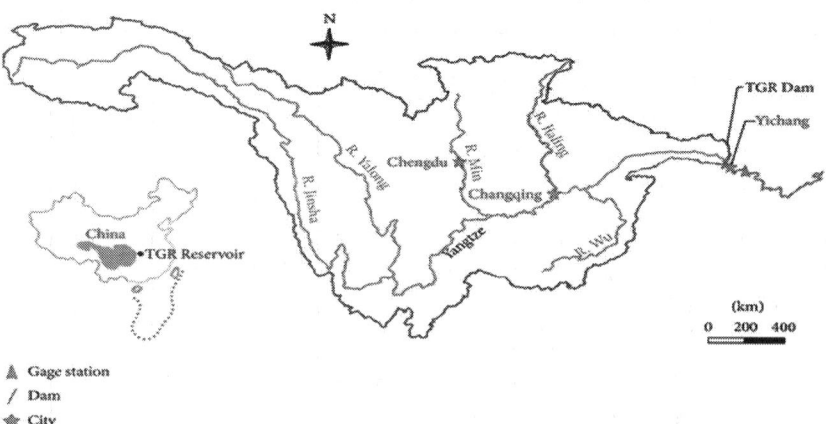

Figure 2: Location of the Three Gorges Reservoir Basin in China.

Several big floods in the Yangtze River basin, including the flood in 1981, have caused serious disasters. Based on the Chinese guidelines for design flood, the flood in 1981 is used as the typical flood to design flood hydrographs of 20-year return period flood (the flood prevention standard for the Yangtze River). Finally, the design flood hydrograph of the TGR, with a return period of 20-year, is used to test the proposed method. The optimal scheduling of the proposed MILP model is compared with the conventional method.

TGR Spillways

For the TGR, there are five types of spillways: turbines, deep outlets, floats outlets, desilting outlets, and surface outlets. Note that the turbines are taken as spillways owing to its capability of releasing flood, and they should be fully open to generate hydropower during the flood events. The numbers of various types of spillways and the code for the formulation are shown in Table 1.

Table 1: Numbers of spillways for various types and the code for the MILP model

Spillway gate	Turbines	Deep outlets	Floats outlets	Desilting outlets	Surface outlets
Number	26	23	2	7	22
Code	1	2–13	14-15	16–22	23–33

Note that the paired deep outlets are denoted as the codes number from 2 to 12, for the consideration of symmetry. For example, code 2 denotes the symmetric deep outlets, 1 and 23. Similarly, the codes from 23 to 33 denote the symmetry paired surface outlets.

The spillways must be fully open or closed for the safety and life span of facilities. Two kinds of spillway constraints should be taken into consideration when scheduling the reservoir system: (1) the potential maximum release for each individual spillway corresponding to specific reservoir storage and (2) the control order of the spillways.

- Relationship between Water Release and Reservoir Storage. The water release of each individual spillway depends on its type and status (open, closed, and scale open). Spillway relationships between reservoir storage and water release are given in Table 2.

Table 2: Water release relationships of spillways

Water level (m)	Reservoir storage (billion m³)	Water releases (m³/s)				Turbines (26 units)	Total release
		Deep outlets	Floats outlets	Desilting outlets	Surface outlets		
		(23 outlets)	**(2 outlets)**	**(7 outlets)**	**(22 outlets)**		
135	12.40	33500	100	2200	/	22100	57900
140	14.70	35800	700	2300	/	23000	61800
155	22.80	41800	3500	/	/	25200	70500
158	24.81	42900	3900	· /	0	25600	72400
160	26.20	43600	4100	/	800	25900	74400
165	30.02	45400	4600	/	5800	25800	81600

- Control Order of Spillways. The spillways should be operated with a specific order, which are described as follows.
- The spillways should be opened in the following order: turbines; deep outlets; floats outlets; desilting outlets; surface outlets. The spillways should be closed in the reversed order.
- The deep outlets, floats outlets, and desilting outlets should be either fully open or fully closed. The partial open is not allowed in the operation.
- The deep outlets and surface outlets should be evenly and symmetrically used, in order that the water release can be distributed evenly along the dams. The spillways should be closed in a reversed order and the concentrated water release at the same location must be prohibited.
- For the floats outlets, floats outlet 2 should be used before the use of the floats outlet 1.
- The desilting outlets are mainly responsible for the sediment releasing and the water release should be avoided in the operations. The water level in the reservoir should be kept below 150 meters if the desilting outlets have to be used for the water releasing.
- Desilting outlets 2 to 6 should be opened earlier and the desilting outlets 1 and 7 can be followed.

The Conventional Operation

Based on the conventional operating rules, the reservoir water release should be kept below 56700 m³/s. That is, the reservoir release is equal to the inflow when the water level is lower than 145 m and the inflow is less than 56700 m³/s; otherwise the water release is equal to 56700 m³/s. When the water level is higher than the maximum flood level (175 m), the water release is equal to the potential maximum water release for the consideration of dam safety. It should be noted that the conventional operating rules are the optimal solution for the model that consists of (1) to (5).

However, this release should be specified to the spillway to satisfy their operation constraints. The allocation is performed by trial and error method and the result is shown in Figure 3. The maximum reservoir storage is 22.09 billion m³.

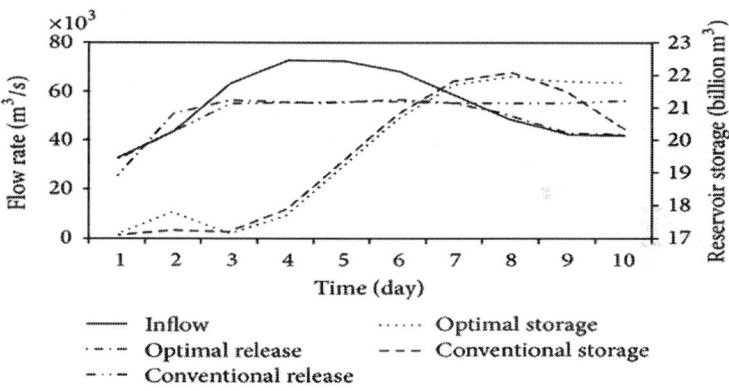

Figure 3: TGR operation for 20-year flood with 1981 type.

Optimal Operation

MILP Model

Since there are five types of spillways, including turbines, deep outlets, floats outlets, desilting outlets, and surface outlets, for the TGR. With the assistance of the binary variables, the water release of each individual spillway can be formulated in piecewise linear relationship between reservoir storage and water release in Section 2. With the objective function of (6) and the spillways constraints, the MILP model has been set up for the TGR finally (see Appendix). In the model, O_{max} and M are set as $100000\,m^3/s$ and 100, respectively, and they are proper for the TGR case.

MILP Solver

The MILP model is resolved by using IBM ILOG CPLEX Optimization Studio (CPLEX) [32], with the interface of the General Algebraic Modeling System (GAMS) [33]. The GAMS is specifically designed for modeling linear, nonlinear, and mixed integer optimization problems. CPLEX is an optimization software package, solving integer programming, linear programming, convex, and nonconvex quadratic programming and so on problems. The CPLEX is accessible through GAMS in this study.

Results of Optimal Scheduling

The optimal scheduling has been found by using the CPLEX solving the MILP model. Table 3 lists the results of the optimal scheduling of 20-year flood. As shown in Table 3, the numbers of spillways of different types opened in different time intervals have already been determined. Furthermore, the status of each individual spillway can also been determined.

Table 3: Optimal scheduling of 20-year flood with 1981 type

Time (day)	Inflow (m³/s)	Storage (billion m³)	Release (m³/s)	Turbine	Deep outlets	Floats outlets	Desilting outlets	Surface outlets
1	32875	17.15	25347	26	1	0	0	0
2	43250	17.80	50626	26	16	0	0	0
3	62875	17.16	56669	26	20	0	0	0
4	72550	17.70	55538	26	19	0	0	0
5	72525	19.17	55240	26	18	0	0	0
6	68475	20.66	56409	26	18	0	0	0
7	58200	21.71	55354	26	17	0	0	0
8	48500	21.95	50136	26	14	0	0	0
9	42400	21.81	42887	26	10	0	0	0
10	42000	21.77	42867	26	10	0	0	0

Since all the turbines are opened and all the floats outlets, desilting outlets, and surface outlets are closed during the whole process of the flood, the descriptions of the statuses of the turbines, floats outlets, desilting outlets, and surface outlets are relatively meaningless. The optimal status of the deep outlets is listed in Table 4.

Table 4: Status of the deep outlets of 20-year flood with 1981 type, where ● and ○ imply that the deep outlet is open and closed, respectively

Time	01	02	03	04	05	06	07	08	09	10	11	12	13	14	15	16	17	18	19	20	21	22	23
1	○	○	○	○	○	○	○	○	○	○	○	●	○	○	○	○	○	○	○	○	○	○	○
2	○	○	○	●	●	●	●	●	●	●	●	●	○	●	●	●	●	●	●	●	○	○	○
3	○	●	●	●	●	●	●	●	●	●	●	○	●	●	●	●	●	●	●	●	●	●	○
4	○	○	●	●	●	●	●	●	●	●	●	●	●	●	●	●	●	●	●	●	●	○	○
5	○	○	●	●	●	●	●	●	●	●	●	○	●	●	●	●	●	●	●	●	○	○	○
6	○	○	●	●	●	●	●	●	●	●	●	○	●	●	●	●	●	●	●	●	●	○	○
7	○	○	○	●	●	●	●	●	●	●	●	●	●	●	●	●	●	●	●	○	○	○	○
8	○	○	○	○	●	●	●	●	●	●	●	○	●	●	●	●	●	●	○	○	○	○	○
9	○	○	○	○	○	○	●	●	●	●	●	○	●	●	●	●	●	●	○	○	○	○	○
10	○	○	○	○	○	○	●	●	●	●	●	○	●	●	●	●	●	○	○	○	○	○	○

As shown in the Table 4, the filled circle implies that the deep outlet is open and the empty circle implies that the deep outlet is closed. It demonstrates the whole process of scheduling of the 20-year flood.

As shown in Figure 3, the optimal scheduling is compared with the conventional method. With the comparison of the results of 20-year flood, the following findings can be observed.(1)The maximum reservoir storage, 21.95 billion m³ in the optimal scheduling, is lower than that in the conventional scheduling 22.09 billion m³, indicating that the optimization is effective. The proposed MILP model provides more available reservoir storage for potential floods. Indeed, it is able to find the global optimum.(2)The maximum water releases of the optimal and conventional method are 56669 m³/s and 56517 m³/s, respectively. These releases are feasible for the downstream safety. Since the optimal operation prereleases more water before the flood peak occurs, it outperforms the conventional operation.(3)Compared with the conventional method, the proposed model specifies the spillways status directly without the allocation using trial and error methods, making the operation more objective. The spillway gates can be easily operated according to the optimal results (Table 4).

CONCLUSIONS

This paper proposes a MILP model to determine the optimal reservoir spillways scheduling. The piecewise linear approximation is used to formulate the relationship between the reservoir storage and water releases for spillways. The control order and symmetry rules of the spillways are described and incorporated into the constraints. Conclusions can be drawn as follows.(1)The optimal scheduling obtained with the MILP model is better than the conventional scheduling in terms of objective function.(2)The optimal scheduling is more advantageous than the conventional scheduling in that the spillways status can be specified directly from the MILP model without water release allocation based on trial and error methods and that the global optimum is ensured.

However, the MILP model is time consuming and the extension of multireservoir systems operation needs further research.

ACKNOWLEDGMENTS

This study was supported by the Program for New Century Excellent Talents in University (NCET-11-0401), the Non-Profit Industry Financial Program of Ministry of Water Resources (201201051), the Central Water Resources Allocation Fee Program (1261430210028), and the National Natural Science Foundation of China (51190094).

REFERENCES

1. S. Guo, H. Zhang, H. Chen, D. Peng, P. Liu, and B. Pang, "A reservoir flood forecasting and control system for China," Hydrological Sciences Journal, vol. 49, no. 6, pp. 959–972, 2004.

2. C. Cheng and K. W. Chau, "Flood control management system for reservoirs," Environmental Modelling and Software, vol. 19, no. 12, pp. 1141–1150, 2004.

3. S. Wang and G. H. Huang, "A two-stage mixed-integer fuzzy programming with interval-valued membership functions approach for flood-diversion planning," Journal of Environmental Management, vol. 117, pp. 208–218, 2013.

4.   W. W.-. Yeh, "Reservoir management and operations models: a state-of-the-art review," Water Resources Research, vol. 21, no. 12, pp. 1797–1818, 1985.

5.   J. W. Labadie, "Optimal operation of multireservoir systems: state-of-the-art review," Journal of Water Resources Planning and Management, vol. 130, no. 2, pp. 93–111, 2004.

6.   D. Karaboga, A. Bagis, and T. Haktanir, "Controlling spillway gates of dams by using fuzzy logic controller with optimum rule number," Applied Soft Computing Journal, vol. 8, no. 1, pp. 232–238, 2008.

7.   C. Wei and N. Hsu, "Optimal tree-based release rules for real-time flood control operations on a multipurpose multireservoir system," Journal of Hydrology, vol. 365, no. 3-4, pp. 213–224, 2009.

8.   A. Bagis and D. Karaboga, "Evolutionary algorithm-based fuzzy PD control of spillway gates of dams,"Journal of the Franklin Institute, vol. 344, no. 8, pp. 1039–1055, 2007.

9.   L. Chang, "Guiding rational reservoir flood operation using penalty-type genetic algorithm," Journal of Hydrology, vol. 354, no. 1–4, pp. 65–74, 2008.

10.   X. Li, S. Guo, P. Liu, and G. Chen, "Dynamic control of flood limited water level for reservoir operation by considering inflow uncertainty," Journal of Hydrology, vol. 391, no. 1-2, pp. 124–132, 2010.

11.   G. Fu, "A fuzzy optimization method for multicriteria decision making: an application to reservoir flood control operation," Expert Systems with Applications, vol. 34, no. 1, pp. 145–149, 2008.

12.   H. Hashemi, J. Bazargan, S. M. Mousavi, and B. Vahdani, "An extended compromise ratio model with an application to reservoir flood control operation under an interval-valued intuitionistic fuzzy environment," Applied Mathematical Modelling, vol. 38, no. 14, pp. 3495–3511, 2014.

13.   A. Karbowski, K. Malinowski, and E. Niewiadomska-Szynkiewicz, "A hybrid analytic/rule-based approach to reservoir system management during flood," Decision Support Systems, vol. 38, no. 4, pp. 599–610, 2005.

14. J. T. Needham, D. W. Watkins Jr., J. R. Lund, and S. K. Nanda, "Linear programming for flood control in the Iowa and Des Moines Rivers," Journal of Water Resources Planning and Management, vol. 126, no. 3, pp. 118–127, 2000.

15. M. R. Norouzi, A. Ahmadi, A. E. Nezhad, and A. Ghaedi, "Mixed integer programming of multi-objective security-constrained hydro/thermal unit commitment," Renewable and Sustainable Energy Reviews, vol. 29, pp. 911–923, 2014.

16. D. W. Watkins, D. J. Jones, and D. T. Ford, "Flood control optimization using mixed-integer programming," in Proceedings of the Flood Control Optimization Using Mixed-Integer Programming Conference, Reston, Va, USA, 1999.

17. P. Liu, T. Nguyen, X. Cai, and X. Jiang, "Finding multiple optimal solutions to optimal load distribution problem in hydropower plant," Energies, vol. 5, no. 5, pp. 1413–1432, 2012.

18. P. Guo, G. H. Huang, and Y. P. Li, "An inexact fuzzy-chance-constrained two-stage mixed-integer linear programming approach for flood diversion planning under multiple uncertainties," Advances in Water Resources, vol. 33, no. 1, pp. 81–91, 2010.

19. C. Ma, J. Lian, and J. Wang, "Short-term optimal operation of Three-gorge and Gezhouba cascade hydropower stations in non-flood season with operation rules from data mining," Energy Conversion and Management, vol. 65, pp. 616–627, 2013.

20. M. Breckpot, O. M. Agudelo, P. Meert, P. Willems, and B. D. Moor, "Flood control of the demer by using model predictive control," Control Engineering Practice, vol. 21, no. 12, pp. 1776–1787, 2013.

21. N. Hsu and C. Wei, "A multipurpose reservoir real-time operation model for flood control during typhoon invasion," Journal of Hydrology, vol. 336, no. 3-4, pp. 282–293, 2007.

22. E. C. Özelkan, Á. Galambosi, E. Fernández-Gaucherand, and L. Duckstein, "Linear quadratic dynamic programming for water reservoir management," Applied Mathematical Modelling, vol. 21, no. 9, pp. 591–598, 1997.

23. X. Fu, A. Li, L. Wang, and C. Ji, "Short-term scheduling of cascade reservoirs using an immune algorithm-based particle swarm

optimization," Computers and Mathematics with Applications, vol. 62, no. 6, pp. 2463–2471, 2011.

24. Q. Zou, J. Zhou, C. Zhou et al., "Fuzzy risk analysis of flood disasters based on diffused-interior-outer-set model," Expert Systems with Applications, vol. 39, no. 6, pp. 6213–6220, 2012.

25. E. Danso-Amoako, M. Scholz, N. Kalimeris, Q. Yang, and J. Shao, "Predicting dam failure risk for sustainable flood retention basins: a generic case study for the wider Greater Manchester area,"Computers, Environment and Urban Systems, vol. 36, no. 5, pp. 423–433, 2012.

26. Y. Ding and S. S. Y. Wang, "Optimal control of flood diversion in watershed using nonlinear optimization," Advances in Water Resources, vol. 44, pp. 30–48, 2012.

27. C. Wei and N. Hsu, "Multireservoir real-time operations for flood control using balanced water level index method," Journal of Environmental Management, vol. 88, no. 4, pp. 1624–1639, 2008.

28. P. Liu, X. Cai, and S. Guo, "Deriving multiple near-optimal solutions to deterministic reservoir operation problems," Water Resources Research, vol. 47, no. 8, article 7208, 2011.

29. Design of Small Dams, United States Department of the Interior, Bureau of Reclamation, 3rd edition, 1987.

30. A. Ashouri, S. S. Fux, M. J. Benz, and L. Guzzella, "Optimal design and operation of building services using mixed-integer linear programming techniques," Energy, vol. 59, pp. 365–376, 2013.

31. P. Luathep, A. Sumalee, W. H. K. Lam, Z. Li, and H. K. Lo, "Global optimization method for mixed transportation network design problem: a mixed-integer linear programming approach," Transportation Research B: Methodological, vol. 45, no. 5, pp. 808–827, 2011.

32. "CPLEX User›s Manual," IBM ILOG CPLEX Optimization Studio,ftp://ftp.software.ibm.com/software/websphere/ilog/docs/optimization/cplex/ps_usrmancplex.pdf.

33. GAMS—A User›s Guide, GAMS Development Corporation, Washington, DC, USA,http://www.gams.com/dd/docs/bigdocs/GAMSUsersGuide.pdf.

Chapter 6

Water Breakthrough Shape Description of Horizontal Wells in Bottom-Water Reservoir

Shijun Huang[1], Baoquan Zeng[2], Fenglan Zhao[3],
Linsong Cheng[1], and Baojian Du[1]

[1]CMOE Key Laboratory of Petroleum Engineering, China University of Petroleum, Beijing 102249, China

[2]Petrochina Research Institute of Petroleum Exploration & Development, Beijing 102249, China

[3]The EOR Center, China University of Petroleum, Beijing 102249, China

ABSTRACT

Horizontal wells have been applied in bottom-water reservoir since their advantages were found on distribution of linear dropdown near wellbore, higher critical production, and more OOIP (original oil in place) controlled. In the paper, one 3D visible physical model of horizontal physical model is designed and built to simulate the water cresting process during the horizontal well producing and find water breakthrough point in homogenous and heterogeneous reservoir with bottom water. Water cresting shape and water cut of horizontal well in between homogenous and heterogeneous reservoir are compared on the base of experiment's result. The water cresting pattern of horizontal well in homogeneous reservoir can be summarized as "central breakthrough, lateral expansion, thorough flooding, and then flank uplifting." Furthermore, a simple analysis model of horizontal well in bottom water reservoir is established and water breakthrough point is analyzed. It can be drawn from the analysis result that whether or not to consider the top and bottom border, breakthrough would be located in the middle of horizontal segment with equal flow velocity distribution.

INTRODUCTION

In the 1990s, with wide application of Hele-Shaw model, 2-D (two-dimensional) physical experiment had been gradually used to investigate performance of horizontal wells in reservoirs with bottom water. Permadi's research showed that water cresting will first present at heel of horizontal segment; then it would gradually move toward to toe [1]; Souza established a new mathematical model to analyse water breakthrough time and flow regime after breakthrough with reservoir parameters, compared with that of numerical simulation by Eclipse, Jiang, Tove Aulie, Francols M, Giger and Wang et al. investigated the water cresting of horizontal well with the mathematical model above and got the similar conclusion [2–6].

Logarithmic pressure distribution of vertical well would result in higher pressure loss near wellbore than that of horizontal well [7]. Due to its particular contraction with reservoir, horizontal well has linear pressure distribution and low pressure loss near wellbore; the corresponding OWC (oil-water contract) would rise with cresting shaped. Therefore the horizontal wells have much lower pressure difference than that of vertical wells with the same production and its critical production is significantly greater than the vertical; on the other hand, the horizontal wells have larger OOIP and swept volume, higher cumulative oil production without water, and longer water free production period; all those would make its wide application in bottom-water reservoir.

At present, investigation on horizontal wells developing bottom water reservoir is not yet systematical due to complex wellbore structure. First, there are few literatures particularly describing water cresting process and detecting breakthrough location, which led to unreasonable completion design and undirected measures of water shutoff and water control; finally it is unfavorable for development of bottom water reservoir [8, 9]. Second, performance and influence of horizontal wells had not been summarized systematically and the corresponding relationship between geology and performance has not yet been established.

In this paper, combined with numerical simulation and 3-D physical simulation, fluid mechanics in porous medium was used to investigate water cresting and breakthrough location, providing consistent theory for reasonable measures of water shutoff and water control.

WATER CRESTING EXPERIMENT WITH HORIZONTAL WELL

After heel-toe pressure difference of horizontal wells was calculated with coupling model [10], majority of reservoir engineers consider that lower pressure make the breakthrough location appear at the heel rather than at the toe. Moreover, P. Permadi's 2-D plate experiment verified this theory with photos of water cresting. Then, a lot of experimental results from Hele-Shaw model showed that heel-toe pressure differences have a significant influence on water cresting. In addition, Q. Jiang, Tove Aulie, Francols M, Giger, and Wang Jialu et al. investigated the water cresting of horizontal wells with similar physical model and got similar conclusion.

With extremely simplified formula; most scholars believe that water cresting would rise asymmetrically, and breakthrough would present at the heel [11]. However, the experimental model and parameters have not been scientifically demonstrated, and the results may be not coordinated with production data.

Firstly, 2-D model can only simulate piston-displacement in heterogeneous reservoir while ellipsoid pressure field cannot be simulated due to walls of device; therefore the water cresting simulated would have a large difference with the one in reservoir.

Secondly, the experiment devices without sand keep distance of two parallel glasses only 0.00625 m in order to increase flow resistance, and the obtained permeability about 32000 μm² (3.2×10^7 mD) is not practical.

If the flow velocity gets to a certain value, both viscous resistance and inertial resistance are generated; meanwhile nonlinear relationship would present between flow velocity and pressure difference, and the flow velocity was defined as critical velocity for Darcy's law. Reynolds criterion was used by Katja Hof to calculate the critical flow velocity [12]:

$$v_c = \frac{3.5\mu\phi^{2/3}}{\rho\sqrt{K}}.$$

(1)

As mentioned in literature the lowest average velocity of experiments is 7.04×10^{-3} m/s, which is much higher than critical velocity of 1.66×10^{-3} m/s from a certain reservoir and is not coordinate with flow velocity in porous media. Without sand, Hele-Shaw mode has extremely high permeability and nearly no flow resistant for bottom water. As a result, generation as well as development of water cresting has no practical significance in that model.

Rather than 2-D model, 3-D physical model had been carried out successively to accurately simulated reservoir parameters, to reflect inflow performance of horizontal wells with different completions, and to better analyze influence factors of well productivity; therefore, a new device was designed to investigate water cresting shape and to detect breakthrough location in Figure 1.

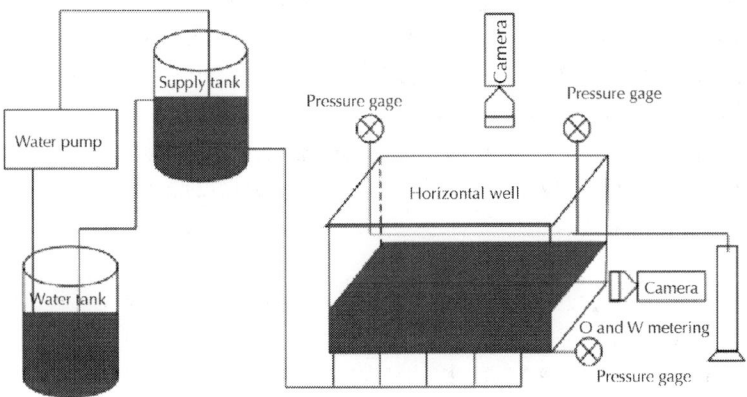

Figure 1: 3-D physical model of horizontal wells in bottom water reservoir.

Water Cresting Formation and Development

As shown in Figure 2, water cresting formation and development can be investigated with pictures at the time of beginning, water breakthrough, and watercut about 40%, 60%, 90%, in which horizontal segment length is 7 cm, 14 cm, and 21 cm separately. And several typical water cresting types of horizontal wells were obtained from experimental results above in homogeneous reservoir, with the same watercut, horizontal segment about 7 cm had much more severe OWC deformation, more remaining oil at two flanks of water cresting, while OWC of horizontal segment about 21 cm is relatively flat.

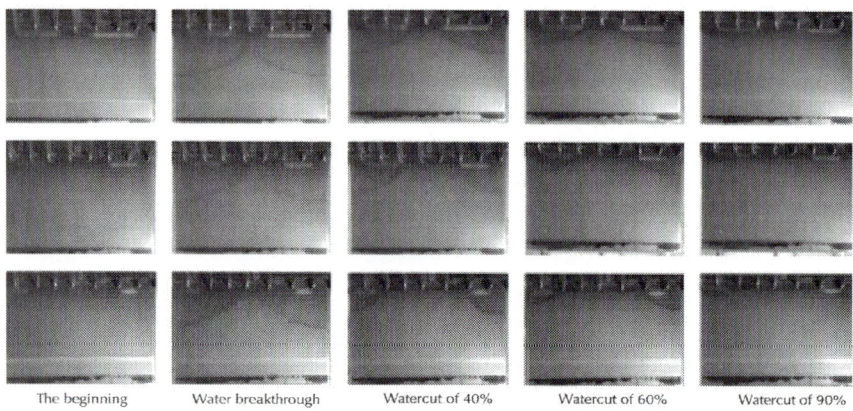

| The beginning | Water breakthrough | Watercut of 40% | Watercut of 60% | Watercut of 90% |

Figure 2: The formation and development of water cresting with different horizontal segment length.

Throughout the whole process, flat OWC will rise steady firstly; then it was gradually deformed into water cresting and breakthrough at the middle of horizontal well; finally the length of water cresting would extended along the horizontal segment from the middle to the toe and the heel until the whole segment was flooded.

Water Cresting Shape in Heterogeneous Reservoirs

Two physical models with different horizontal segment length were designed (Figure 3) to analyze water cresting shape in heterogeneous reservoir. In each model, there are two areas with permeability of $10.11\,\mu m^2$ and $2.33\,\mu m^2$ separately along horizontal segment, and permeability contrast is almost 4 times. In the first one, the horizontal segment about 14 cm was perforated in both areas; in the second one, the horizontal segment length of 7 cm was only located and performed in low permeability area.

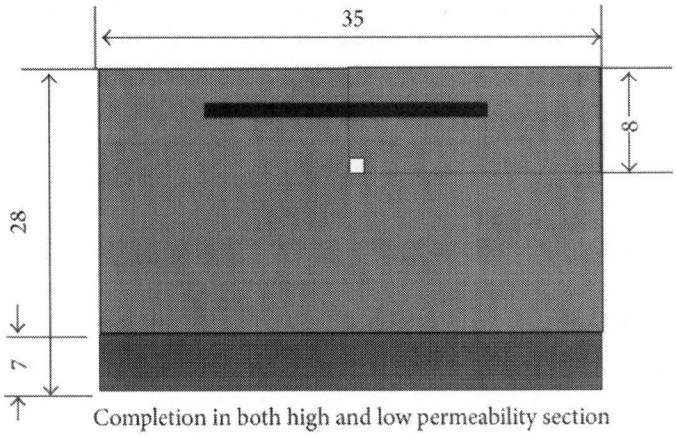

Completion in both high and low permeability section

(a)

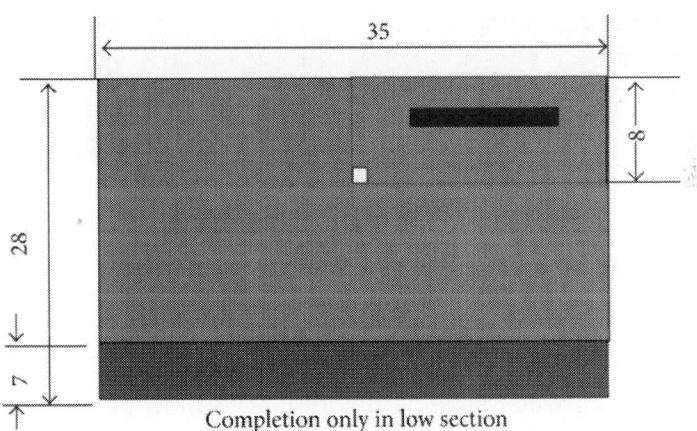

Completion only in low section

(b)

Figure 3: Physical models with different horizontal segment length.

In the first model (Figure 4), bottom water was advanced so sharply in high permeability area that lower permeability area would not be flooded and be rich of banded remaining oil, which is similar to "eaves oil" with bowl shaped because there is nearly no pressure difference as well as no production in low permeability area; therefore this is the main reason why production of horizontal well in reservoir is much lower than theoretical one. The water cresting process can be described as follows: tilted advancing → high permeability raise → high permeability flooded → flank uplift → no flooding in low permeability; experimental results showed that if permeability contrast is larger than a certain value, there would be much remaining oil in low permeability area and "eaves oil," which would become new targets of potential tapping in high watercut stage.

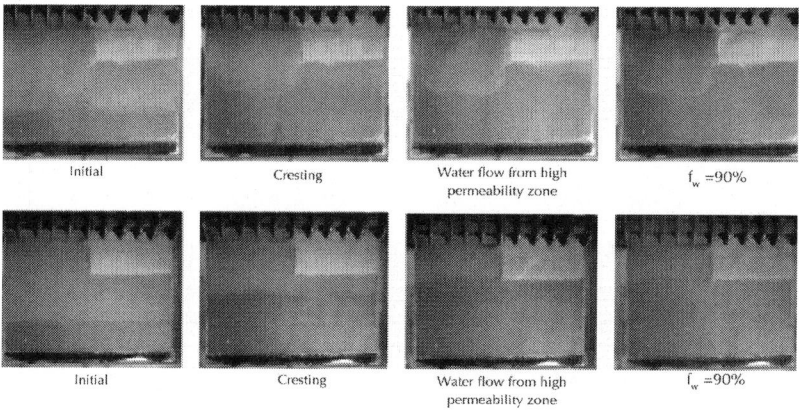

Figure 4: The formation and development of water cresting with low and high permeability section.

In the second model (Figure 4), the bottom water always advanced steadily without water cresting before it got to low permeability area. The low permeability area was flooded without banded remaining oil in its lower part. However, throughout the whole production high permeability area had no water flooding while low permeability area had been flooded with scattered remaining oil. And that phenomenon is much coordinated with displacement characteristics in low permeability reservoir.

Permeability heterogeneity had distinguished impact on watercut, rising step by step, and the number of steps was coordinate with number of nonconsecutive high permeability areas covered by horizontal segment.

To better investigate coupling effect between flow in reservoir and variable mass flow in wellbore, a new model was established to analyze the water cresting with different parameters.

THEORETICAL INVESTIGATION ON WATER CRESTING

Horizontal Well with (2n+1) Perforation

As shown in Figure 5, coordinates of perforations are ,(0,0),($\pm$a,0),($\pm$2a,0),...,($\pm$na,0) and longitudinal velocity at any point is as follows:

$$v_y = v_{y_0} + v_{y_1} + v_{y_{1'}} + \cdots + v_{y_n} + v_{y_{n'}}$$

$$= -\frac{\partial \Phi_0}{\partial y} - \frac{\partial \Phi_1}{\partial y} - \frac{\partial \Phi_{1'}}{\partial y} - \cdots - \frac{\partial \Phi_n}{\partial y} - \frac{\partial \Phi_{n'}}{\partial y}$$

$$= \frac{q}{\pi} y \left[\frac{1}{x^2 + y^2} + \frac{1}{(x + a)^2 + y^2} + \frac{1}{(x - a)^2 + y^2} \right.$$

$$\left. + \cdots + \frac{1}{(x + na)^2 + y^2} + \frac{1}{(x - na)^2 + y^2} \right],$$

$$v_y \Big|_{x=0, y=y_0} = \frac{q}{\pi} y_0 \left[\frac{1}{y_0^2} + \frac{2}{a^2 + y_0^2} + \cdots + \frac{2}{(na)^2 + y_0^2} \right]$$

$$= \frac{q}{\pi} y_0 \left[\sum_{i=0}^{n} \frac{1}{i^2 a^2 + y_0^2} + \sum_{i=1}^{n} \frac{1}{i^2 a^2 + y_0^2} \right],$$

$$(2)$$

$$v_y\Big|_{x=ja,y=y_0} = \frac{q}{\pi} y_0 \left[\frac{1}{(ja)^2 + y_0^2} + \frac{1}{(j+1)^2 a^2 + y_0^2} \right.$$

$$+ \frac{1}{(j-1)^2 a^2 + y_0^2}$$

$$+ \cdots + \frac{1}{(n+j)^2 a^2 + y_0^2}$$

$$\left. + \frac{1}{(n-j)^2 a^2 + y_0^2} \right]$$

$$= \frac{q}{\pi} y_0 \left[\sum_{i=0}^{n} \frac{1}{i^2 a^2 + y_0^2} + \sum_{i=n}^{2n+1} \frac{1}{i^2 a^2 + y_0^2} \right],$$

$$v_y\Big|_{x=0,y=y_0} > v_y\Big|_{x=ja,y=y_0}.$$

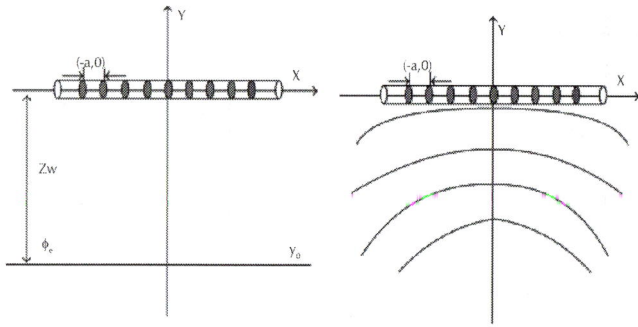

Figure 5: Crest schematic chart of horizontal wells with (2n+1) perforations.

The longitudinal velocity of middle perforation is always the highest, and the water cresting shape was shown in Figure 5 (right).

When $n \to +\infty$ and $a \to 0$, it was equal to a horizontal well with barefoot completion, and breakthrough point would be located in the middle of horizontal segment considering no pressure loss.

Water Cresting with Top and Bottom Boundary

Physical model shown in Figure 6 had a closed upper boundary and lower boundary with constant pressure. And the horizontal well can be considered as a row of (2n+1) wells with the same flow velocity. Based on image theory, in infinite space each perforation can be mapped to well array with an interaction of two injections and two productions. All wells can be divided into four categories: (1) injection well with coordination of $(ia, 2h + 4nh + Zw)$;(2) injection well with coordination of $(ia, 4nh - Zw)$;(3) production well with coordination of $(ia, 2h + 4nh - Zw)$, and (4) production well with coordination of $(ia, 4nh + Zw)$, where , $i = 0, \pm1, \pm2, \ldots\ldots, \pm n$,

$$\Phi(x, z)$$

$$= \frac{q}{2}$$

$$\times \sum_{-\infty}^{+\infty} \ln\left(\left(\left[(x - ia)^2 + (Z - 2h - 4nh + Zw)^2 \right] \right. \right.$$

$$\times \left[(x - ia)^2 + (Z - 4nh - Zw)^2 \right] \right)$$

$$\times \left(\left[(x - ia)^2 + (Z - 2h - 4nh - Zw)^2 \right] \right.$$

$$\left. \times \left[(x - ia)^2 + (Z - 4nh + Zw)^2 \right] \right)^{-1} \right)$$

$$+ C. \tag{3}$$

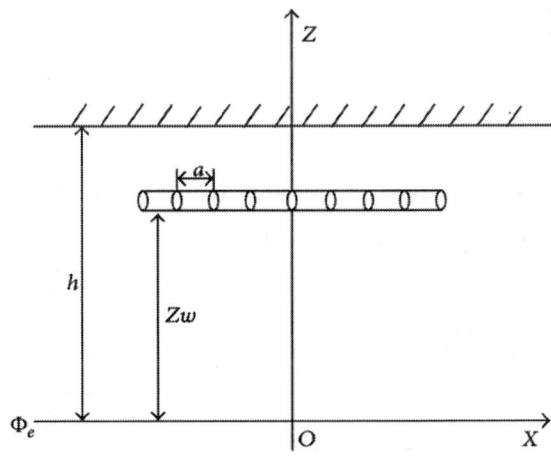

Figure 6: Physical model of horizontal wells in bottom water reservoir (with upper and lower boundaries).

With Behçet formula, (3) can be transformed into

$\Phi(x, z)$

$$
= \frac{q}{2} \ln \left(\left(\left[ch\frac{\pi(x-ia)}{2h} + \cos\frac{\pi(Z+Zw)}{2h} \right] \right. \right.
$$

$$
\times \left[ch\frac{\pi(x-ia)}{2h} - \cos\frac{\pi(Z-Zw)}{2h} \right] \right)
$$

$$
\times \left(\left[ch\frac{\pi(x-ia)}{2h} + \cos\frac{\pi(Z-Zw)}{2h} \right] \right.
$$

$$
\left. \left. \times \left[ch\frac{\pi(x-ia)}{2h} - \cos\frac{\pi(Z+Zw)}{2h} \right] \right)^{-1} \right)
$$

$+ C.$

(4)

Therefore, longitudinal velocity would be

Vz

$$
= -\frac{\partial \Phi(x, z)}{\partial z}
$$

$$
= \frac{\pi q}{2h}
$$

$$
\times \left[\sin\frac{\pi(Z+Zw)}{2h} \right.
$$

$$
\cdot \frac{ch((\pi(x-ia))/2h)}{ch^2((\pi(x-ia))/2h) - \cos^2((\pi(Z+Zw))/2h)}
$$

$$
+ \sin\frac{\pi(Zw-Z)}{2h}
$$

$$
\left. \cdot \frac{ch((\pi(x-ia))/2h)}{ch^2((\pi(x-ia))/2h) - \cos^2((\pi(Zw-Z))/2h)} \right] .
$$

(5)

According to superposition principle, the longitudinal velocity of point (x,z) superimposed by all perforations can be

$$Vz$$

$$= \frac{\pi q}{2h}$$

$$\times \left[\sin \frac{\pi (Z + Zw)}{2h} \right.$$

$$\cdot \sum_{-n}^{n} \frac{ch\left((\pi (x - ia))/2h\right)}{ch^2\left((\pi (x - ia))/2h\right) - \cos^2\left((\pi (Z + Zw))/2h\right)}$$

$$+ \sin \frac{\pi (Zw - Z)}{2h}$$

$$\left. \cdot \sum_{-n}^{n} \frac{ch\left((\pi (x - ia))/2h\right)}{ch^2\left((\pi (x - ia))/2h\right) - \cos^2\left((\pi (Zw - Z))/2h\right)} \right],$$

$$R_1(x) = \frac{ch\left((\pi (x - ia))/2h\right)}{ch^2\left((\pi (x - ia))/2h\right) - \cos^2\left((\pi (Z + Zw))/2h\right)},$$

$$R_2(x) = \frac{ch\left((\pi (x - ia))/2h\right)}{ch^2\left((\pi (x - ia))/2h\right) - \cos^2\left((\pi (Zw - Z))/2h\right)}.$$

$$(6)$$

It could be proved that R is monotone decreasing function of x:

$$Vz|_{(ma,z)} - Vz|_{(0,z)}$$

$$= \frac{\pi q}{2h} \left\{ \sin \frac{\pi (Z + Zw)}{2h} \right.$$

$$\times \left[R_1(na + |m|a) \right.$$

$$- R_1(na) + R_1(na + |m|a - a)$$

$$- R_1(na - a) + \cdots + R_1(na + a)$$

$$\left. - R_1(na - |m|a + a) \right]$$

$$+ \sin \frac{\pi (Zw - Z)}{2h}$$

$$\times \left[R_2(na + |m|a) - R_2(na) \right.$$

$$+ R_2(na + |m|a - a) - R_2(na - a)$$

$$+ \cdots + R_2(na + a)$$

$$\left. \left. - R_2(na - |m|a + a) \right] \right\}.$$

$$(7)$$

While $\sin\big((\pi(Z+Zw))/2h\big)>0, \sin\big((\pi(Zw-Z))/2h\big)>0,$ and $Vz\big|_{(ma,z)}-Vz\big|_{(0,z)}<0$, therefore, the middle one would have the highest longitudinal velocity and earliest water breakthrough with equal velocity at any perforation.

WATER CRESTING WITH INHOMOGENEOUS FLOW WITH TOP AND BOTTOM BOUNDARY

Due to influence of supply area, permeability, reservoir damage, perforation density, flow velocity at any point of horizontal segment is usually unequal. Meanwhile the water cresting shape was carried out with different flow velocity at different nodes.

For different flow velocity at different perforation, (7) would be

$$
\begin{aligned}
Vz &= \frac{\pi}{2h} \\
&\times \Bigg[\sin\frac{\pi\,(Z+Z_w)}{2h} \\
&\quad \cdot \sum_{-n}^{n} q_i \frac{ch\,((\pi\,(x-ia))/2h)}{ch^2\,((\pi\,(x-ia))/2h)-\cos^2\,((\pi\,(Z+Zw))/2h)} \\
&\quad + \sin\frac{\pi\,(Zw-Z)}{2h} \\
&\quad \cdot \sum_{-n}^{n} q_i \frac{ch\,((\pi\,(x-ia))/2h)}{ch^2\,((\pi\,(x-ia))/2h)-\cos^2\,((\pi\,(Zw-Z))/2h)} \Bigg],
\end{aligned}
$$

(8)

where q_i is the flow velocity of perforation i.

In this model, inputting node number and flow velocity of each node, longitudinal velocity of any point below perforations could be calculated with equal (7), and the results would be carried out with a certain interval, until the highest point of water cresting reached the horizontal segment, and OWC at different time will form water resting.

As a result, the water cresting can be calculated with variable nodes and arbitrary flow velocity distribution considering the top and bottom boundaries.

Supposed that, reservoir thickness—30 m, height of water avoidance—25 m, horizontal segment length—300 m with 61 nodes evenly. In (x,z) coordinate plane, calculated with equal (8), the longitudinal velocity at any point is the sum of z partial derivatives for potential superimposed and longitudinal displacement is product of longitudinal velocity and time step. Figures 1–8 showed the water cresting shape with different flow velocity distribution. The production of each node in each fig were $[1,1,......,1],[4,1,1,....,1,1,4],[3,2,1,1,....,1,1,2,3]$, respectively.

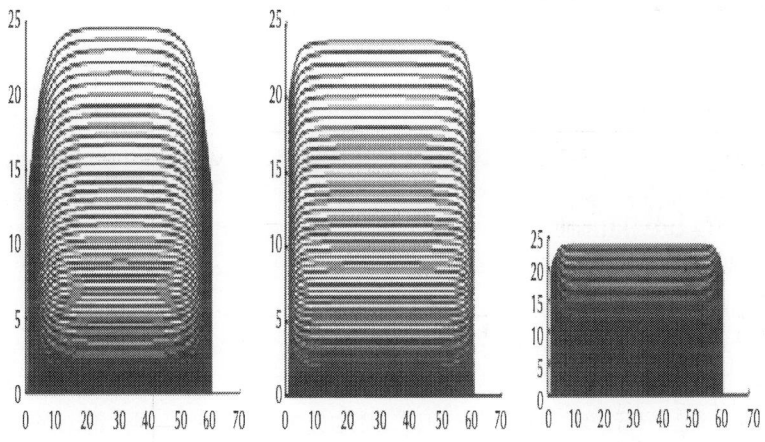

Figure 7: Crest rising shapes with different flow velocity distributions.

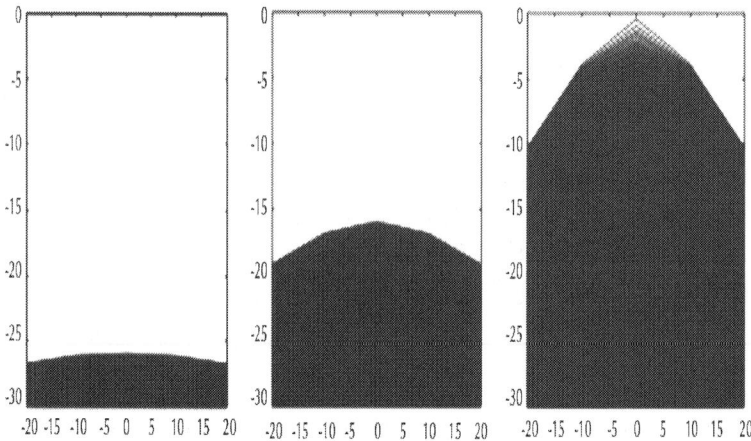

Figure 8: Crest rising shapes with equal inflow velocity.

Calculated results (Figure 7) showed that breakthrough would firstly present in the middle of horizontal well with equal flow velocity even though the velocity at heel or at toe is 4 times that of other points.

In order to investigate water cresting shape with different velocity distribution, the whole wellbore was simplified as five perforations supposing breakthrough time as T and taking water cresting shape into comparison at , T/5,3T/5, and T.

When $[n_1, n_2, n_3, n_4, n_5]$ = [1,1,1,1,1], keeping the dimensionless total flow rate of 305, the breakthrough time was 1706Dt, and water cresting shape was showed in Figure 8.

When $[n_1, n_2, n_3, n_4, n_5]$ = [10,8,5,2,1], namely, decreasing flow velocity, the dimensionless breakthrough time was 1445Dt, and water cresting shape was as shown in Figure 9.

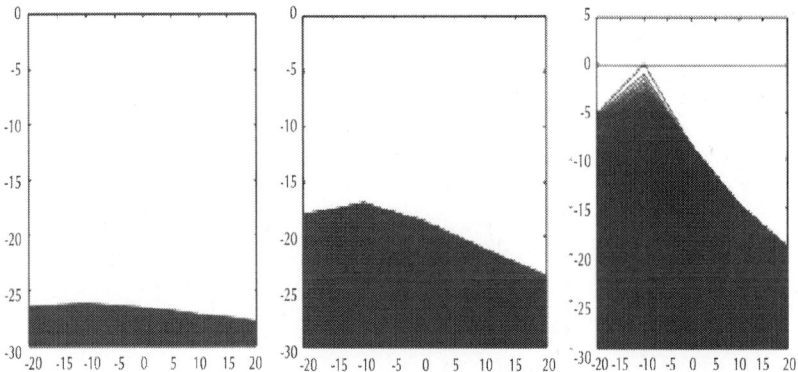

Figure 9: Crest rising shapes with declining inflow velocity from heel to toe.

The water cresting (Figure 9) showed that breakthrough would usually locate in the middle of horizontal segment rather than the point with high flow velocity for superposed pressure field.

When $[n_1, n_2, n_3, n_4, n_5] = [1,2,5,2,1]$, namely, central high-flank low, the dimensionless breakthrough time was 1371Dt, and water cresting shape was as in Figure 10.

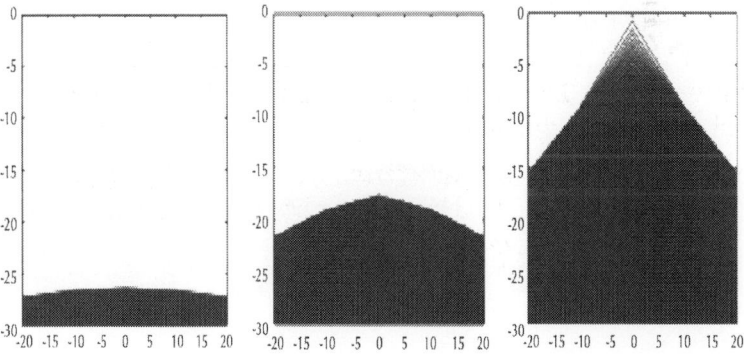

Figure 10: Crest rising shapes with inflow velocity of [1,2,5,2,1] from heel to toe.

When $[n_1, n_2, n_3, n_4, n_5] = [5,2,1,2,5]$, namely, central low-flank high, the dimensionless breakthrough time is 1951Dt, and water cresting shape was showed in Figure 11.

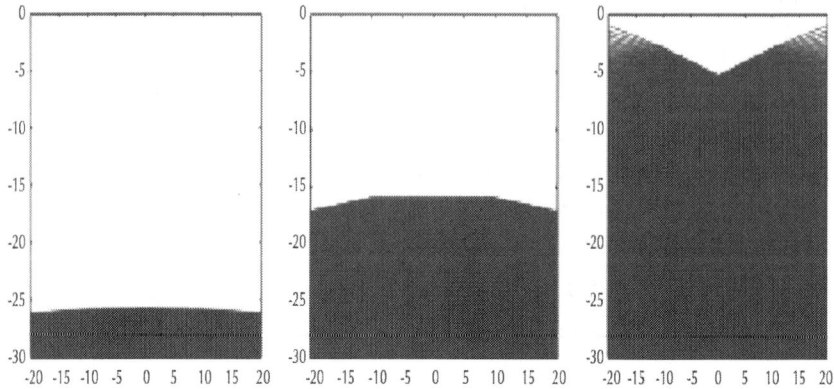

Figure 11: Crest rising shapes with inflow velocity of [5,2,1,2,5] from heel to toe.

When $[n_1, n_2, n_3, n_4, n_5]$ = [25,1,5,1,10], namely, high-low-secondary high, the dimensionless breakthrough time is 1511Dt, and water cresting shape was shown in Figure 12.

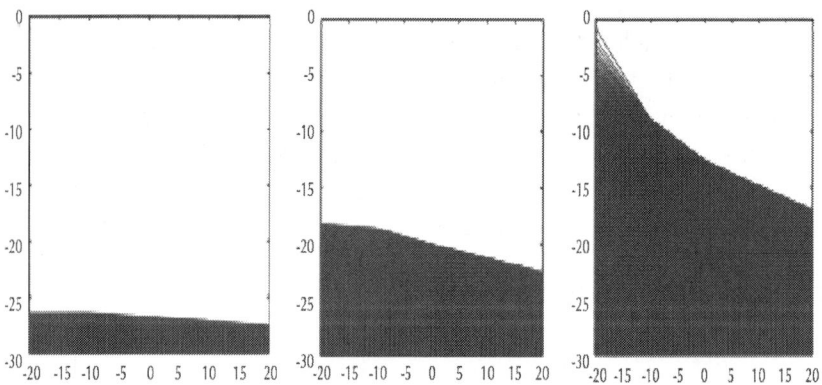

Figure 12: Crest rising shapes with inflow velocity of [25,1,5,1,10] from heel to toe.

With comparison of flow velocity distribution above, conclusions were obtained.

- For horizontal wells in bottom water reservoir, breakthrough would be usually located in the middle of horizontal segment rather than the point with high flow velocity.
- If its middle part was located in high permeability area and was perforated, horizontal well would have shorter water free production period and smaller water free production. Therefore we should pay much more attention to completion and water shutoff.
- Permeability distribution pattern of "central high-flank low" would result in longer water free period and more production without water.

CONCLUSION

- With 3-D visible physical experiments, water cresting in homogeneous reservoir was "central breakthrough → lateral expansion→ thorough flooding → flank uplifting."
- All water cresting shapes are almost similar at the same watercut for horizontal wells with different length. In preceding development, higher drawdown would result in steeper water cresting and much more severe fingering. While watercut > 90%, the whole OWC would arrive at horizontal well.
- Whether or not to consider the top and bottom border, breakthrough would be located in the middle of horizontal segment with equal flow velocity distribution. To some extent, velocity distribution will have some influence on watercut trend, even though the velocity at heel or at toe is 4 times that of other points; breakthrough point would be still located in the middle of horizontal segment with rich remaining oil in flanks of water cresting. Therefore, for horizontal wells in the bottom water reservoir, bottom water will breakthrough not at the point with higher flow velocity but at the middle of horizontal segment for superposed pressure field.

REFERENCES

1. P. Permadi, R. L. Lee, and R. S. T. Kartoatmodjo, "Behavior of water cresting under horizontal well," SPE30374, 1995.

2. A. L. S. Souza, S. Arbabi, and K. Aziz, "A practical procedure to predict cresting behavior in horizontal well," SPE39063, 1997.

3. Q. Jiang and R. M. Butler, "Experimental and numerical modelling of bottom water coning to a horizontal well," Journal of Canadian Petroleum Technology, vol. 37, no. 10, pp. 82–91, 1998.

4. T. Aulie, H. Asheim, P. Oudeman, and P. Laboratorium, "Experimental investigation of cresting and critical flow rate of horizontal wells," SPE26639, 1995.

5. F. M. Giger, "Analytic two-dimensional models of water cresting before breakthrough for horizontal wells," SPE Reservoir Engineering, vol. 4, no. 4, pp. 409–15378, 1989.

6. J.-L. Wang, Y.-Z. Liu, and R.-Y. Jiang, "2-D physical modeling of water coning of horizontal well production in bottom water driving reservoirs," Petroleum Exploration and Development, vol. 34, no. 5, pp. 590–593, 2007.

7. W. Renfu, Horizontal Well Production Technology of Different Type Reservoir in China, Petroleum Industry Press, Beijing, China, 1998.

8. L. Hongshan and W. Shuqiang, "Study on variable density perforation tech for bottom-water reservoir of horizontal well," Well Testing, vol. 17, no. 3, pp. 42–77, 2008.

9. M. Hongxia, C. Dechun, H. Huirong, et al., "Optimization of well completion project for the selectively perforated horizontal well," Petroleum Geology and Recovery Efficiency, vol. 14, no. 5, pp. 84–87, 2007.

10. B. J. Dikken, "Pressure drop in horizontal wells and its effect on their production performance," SPE 19824, 1989.

11. C. Linsong, L. Zhaoxin, and Z. Lihua, "Reservoir engineering problem of horizontal wells coning in bottom-water driven reservoir," Journal of the University of Petroleum, vol. 18, no. 4, pp. 43–46, 1994.

12. Y. Shenglai, Petrophsics, Petroleum Industry Press, Beijing, China, 2004.

Chapter 7

Visualization Requirements of Engineers for Risk Assessment of Embankment Dams

Varun Kasireddy[1], Semiha Ergan[1], Burcu Akinci[1], and Nur Sila Gulgec[2]

[1]Department of Civil and Environmental Engineering, Carnegie Mellon University, 5000 Forbes Ave., Pittsburgh 15213, PA, USA

[2]Department of Civil and Environmental Engineering, Lehigh University, 13 East Packer Avenue, Bethlehem 18015-3176, PA, USA

ABSTRACT

Background

Aging infrastructure in the US has gained quite a bit of attention in the past decade. Being one type of a critical infrastructure, embankment dams in the US require significant investment to upgrade the deteriorated parts. Due to limited budgets, understanding the behavior of structures over time through risk assessment is essential to prioritize dams. During the risk assessment for embankment dams, engineers utilize current and historical data from the design, construction, and operation phases of these structures. The challenge is that during risk assessment, various engineers from different disciplines (e.g., geotechnical, hydraulics) come together, and how they would like to visualize the available datasets changes based on the discipline-specific analyses they need to perform. The objective of this research study is to understand the discipline-specific visualization needs of engineers from US Army Corps of Engineers (USACE) who are involved in risk assessment of embankment dams when they deal with large set of data accumulated since the inception of dams.

Methods

The requirements were identified through a three-phased research approach including interviews with engineers who are regularly involved in risk assessment processes, a card game and review of standards and published work on risk assessment of embankment dams.

Results

This paper provides the findings of research conducted with engineers coming from different disciplines within USACE. Findings comprise discipline-specific visualization requirements of engineers for viewing large datasets, containing static data (e.g. design information) and time-series data (e.g. piezometer data, monument measurements etc.), accumulated since the inception of dams.

Conclusions

The findings suggest that the visualization of the dam layout, components and geometry within 3D settings overlaid with sensor data (which could be queried based on engineers' discipline-specific needs) and data analytics results provide a better flexibility to engineers to understand the risk associated with potential failure modes.

BACKGROUND

Embankment dams, particularly, the aging ones are prone to failure with progressing time. Various types of failures, including internal erosion, sliding due to loading and overtopping, exist for an embankment dam. Many dams in the US have already received a "poor" rating as per the grade card released by American Society for Civil Engineers (ASCE) recently (ASCE [2013]). Most importantly, these dams are an integral part of a prospering economy, and directly concern the lives of a large percentage of population living nearby. To repair and rehabilitate all of those dams are simply not possible due to budget constraints, and hence dams that require immediate remedial actions need to be identified and prioritized. One practical approach for this prioritization is through risk assessment, which includes the assessment of these dams periodically for the level of risk of failure and the magnitude of economic and life causalities associated with such a failure, and act accordingly.

Risk assessment process is an interdisciplinary process and involves engineers of various disciplines like Geotechnical Engineering (GT), Geology (GE), Hydraulic Engineering and Hydrology (H&H), and Structural Engineering (SE). Also, risk assessment activities are typically carried out in different frequencies and granularities. Examples include daily monitoring, which is performed on the daily data collected on the dam to detect changes in readings overtime; periodic inspection (PI), which is conducted every five years in a detailed manner including historical data, and periodic assessment (PA), which is conducted every ten years with interdisciplinary parties. Currently, during these sessions, the multi-disciplinary team of engineers has access to different types of information, such as design, construction and operation information and accesses them through digital or hard copy documents. Collecting the required information and processing/analysing the document based information are resource and time intensive (Shaffner[2011]).

Unique challenges that engineers face during risk assessment include (a) bringing a spatial context to the sensed data from piezometers, inclinometers, survey monuments and weirs, (b) understanding the behaviour of dams over time by correlating several parameters about dams (e.g., evaluating pool elevations with respect to piezometer readings, piezometer readings with respect to their station locations, piezometer tip elevations with respect to soil layers etc.). While data collection and processing efforts are preliminary data stages, it is the data visualization stage that plays a vital role in understanding the valuable information concealed inside the data. As data can be represented in different forms, and stored in multiple formats, it is important to understand which form is the most useful for the end users of the data, i.e., dam engineers in this case, to aid in the risk assessment process. For this purpose, it is necessary to identify the engineer's visualization requirements.

Engineers develop various artefacts to keep track of the correlations in mind, such as correlation plots, cross section layouts, piezometer locations on a plan view, lithology plans showing bore-hole locations and properties. Current tools and artefacts used by engineers do not enable them to perceive the data and correlations between them through views that can be generated flexibly based on how the engineers would like to look at the data. The artefacts are static and are not always capable of correlating the parameters at a glance (USACE [2012]). Likewise, our initial interactions with engineers during a risk assessment session showed that the visualization requirements and corresponding modes of visualization vary as per the background of an engineer. For example, geotechnical engineers require to look at how different rock types are spatially distributed over the dam site and laboratory rock tests reflected as such. On the other hand, geologists intend to look at the same data in a layer-wise manner, and prefer to be able to turn on/off different rock-type layers within the same 2D/3D visualization window. Consequently, this mandates the requirement of a flexible visualization paradigm to ensure effective and efficient perception and comprehension of the data.

Within the context of this paper, the authors provide the details of the findings on identification of discipline specific visualization requirements of engineers needed during risk assessment of embankment dams. The authors describe the challenges of present applications (see section The challenges of current applications), related background research (see section Background research), detail the three-pronged research methodology adopted in this study (see section Methods) and give details of the findings (see section Results and discussion). The identified visualization requirements in this domain are the main contributions of this paper, however the research methodology can be repeated to identify similar requirements in other decision domains. The paper concludes with recommendations and possible future directions.

The Challenges of Current Applications

Current risk assessment procedure involves various engineering disciplines and several different types of documentation. Therefore, the effort is both, time and resource intensive as every engineer involved in the process needs to study the excessive documentation about the dam for weeks and repeat the same efforts in every risk assessment exercise for each dam. There are two main challenges that engineers deal with, as detailed below.

Challenge 1: Time and Resource Intensive Effort to Get a Holistic Understanding of the Dam, Its Behaviour and Its Current Condition

Once the corresponding documents for a dam are collected, a 3–4 week time-frame is then allocated to do the assessment and get a complete view of the dam and its behaviour over time corresponding to the failure modes of that dam. Engineers need to study the documentation and also generate various artefacts to get a holistic view, and some of these artefacts generated for analysing a dam are illustrated in Figures 1, 2, 3 and 4. For instance, in order to understand if any piezometers have anomalies, they use time-series plots (as shown in Figure 1) where they plot the piezometer readings with respect to pool elevations. The interpretation of sensor readings depends on their location with respect to upstream or downstream of the dam, as well as in which soil layer the tip of each piezometer resides. Hence, in addition to the time series plots, they also look at plans that mark piezometer locations on the topography map as well as drawings that show cross sections of the dam with soil layers and piezometer tip elevations (Figures 2 and 3). Since several piezometers exist on a dam body, they generate the same plots for every station on the dam and for each piezometer. When there is large number of instruments on a dam, it becomes a challenge to relate all these artefacts to each other to understand the behaviour of the dam through data collected by the instruments. To help in that process, sometimes engineers also generate transparent physical mock-ups of the dams to keep track of the instrumentation (Figure 4).

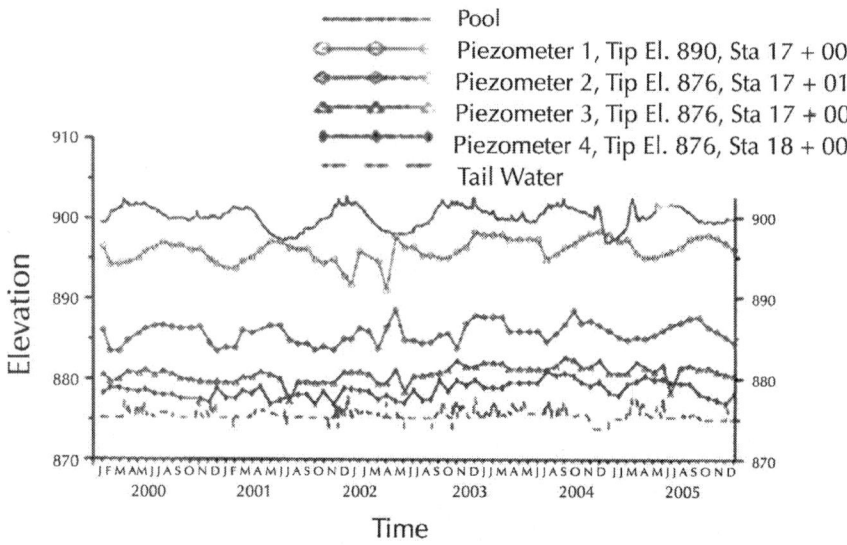

Figure 1: A time series plot showing the piezometer readings with respect to pool elevations.

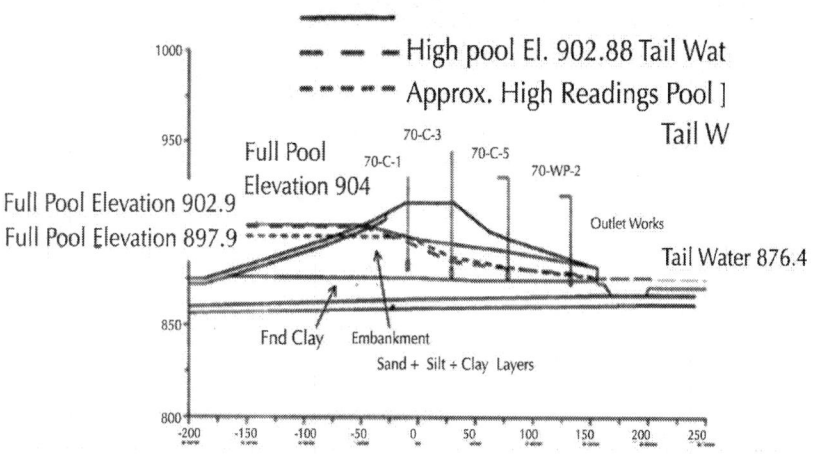

Figure 2: Cross section of a dam showing soil layers and pool elevations.

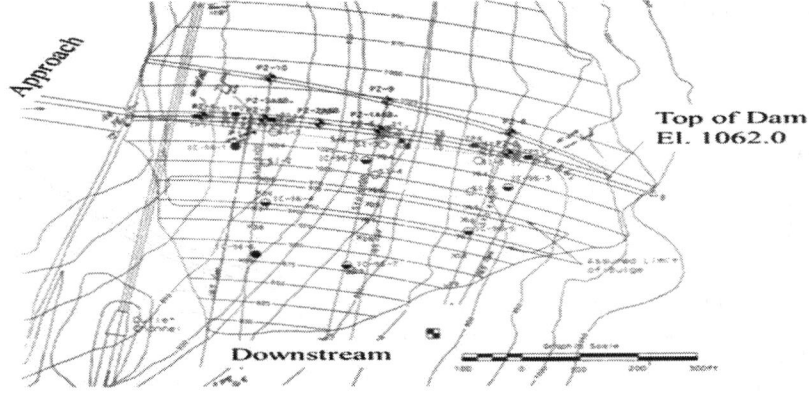

Figure 3: Plan view of piezometer locations.

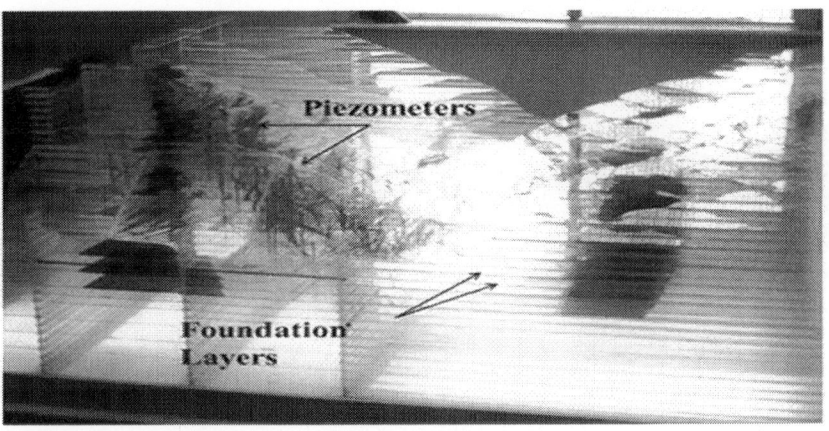

Figure 4: A transparent physical mock-up of a dam (Shaffner [2011]).

Challenge 2: Bringing a Spatial Context to the Sensed Data from Instrumentation

Keeping every information and correlations between them in mind is impossible, given the fact that large number of instruments is used on a given dam and such instrumentation keeps providing data, at the minimum, on a daily basis. For example, a piezometer being read weekly from construction in 1944 through present (2013) would have over 3,500 data points. For a dam with 20 piezometers, it translates to over 70,000 piezometer data points. Meanwhile, a piezometer being read daily from construction in 1944 through present (2013) alone would have over 25,000 data points. Hence, bringing a spatial context to the instrumentation data and correlating them for both visualization and data analysis purposes is essential.

The first step towards such an approach is to identify what information should a shared repository contain. In addition to that, our observations with engineers suggested that engineers from different engineering disciplines want to look at the data from different views and this necessitates that requirements be identified based on engineering discipline. Table 1 is just an example of different visual requirements for looking at reservoir and tail water levels. It is evident that discipline specific requirements should be identified so that the developed repository can support the decisions of all disciplines involved in the risk assessment process. In summary, current applications to assess risk for embankment dams contain challenges of a) not capable of understanding whole behaviour of dam over time, b) visualizing instrumentation data with spatial context, c) lack of ways to generate custom views that support how engineers would like to see the existing data for effective risk assessment. Therefore, this study targets the foundational work required to enable discipline specific visualization, and presents the discipline specific visualization requirements of risk assessors.

Table 1: An example collage of visualizing reservoir and tail water levels based on engineering disciplines

Discipline	Reservoir and tail water level
Geotechnical Engineers	*Would like to see the top 10 events that have occurred and study the differential rating & performance for these events.*
Geologists	*Would like to graphically see water levels per cross-section with piezometer readings embedded.*
Hydrologists	*Would like to visualize inflow-volume-duration-frequency curves (1–7 day computed probabilities).*

Structural Engineers	*Would like to see hydrologic loading data for coincident pools for seismic Probable Mass Function's.*

Kasireddy et al.

Kasireddy et al. Visualization in Engineering 2015 3:1, doi:10.1186/s40327-014-0014-y

Background Research

Several studies in the literature have been done in relation to usage of various forms of visualization to aid the dam risk assessment process. Harnessing different modes of visualization, i.e., 3D and 4D, to present different types of information from disparate sources enhances the ability to absorb the content, as well as the ease of its access, when required (Pantea et al.[2011]). Some of the studies might be grouped such as multi-dimensional visualization, Geographic Information Systems (GIS) and real-time monitoring applications.

Multi-dimensional visualization techniques, which include 3-D and 4-D analyses, enable better understanding, access and presentation of the integrated information from different type of sources (Cross et al. [2005]). 3-D modelling techniques have been widely used to display volumetric characteristics of a given structure, such as surface mapping, surface hydrology, and groundwater levels (Glynn et al. [2011]); the piezometer and water levels (Spencer et al. [2010]) and water surface variations during flood (Lai et al. [2011]). In addition, there are various studies about the modelling of the geometric surfaces and 3D layers that carry the lithological and hydraulic level characteristics (Dominguez-Acosta et al. [2004]) and associated functionalities, such as saving, rotation, zoom, cut and slice (Xi-juan et al. [2010]). Another multi-dimensional technique is 4-D modelling that gives insight for engineers about the behaviour across time. These studies help to analyse the current conditions and guess the possible future actions by displaying time variant behaviours (Brindley et al. [2006]).

Geographic Information Systems (GIS) have been widely used in approaches to remedy information access about dams. In these studies, different data or their consequences can be displayed over the geologic maps. Various applications include (a) modelling of dam performances, such as levee performance and the failure modes and their visualization (Serre et al. [2008]), (b) flood damage estimations (Qi and Altinakar [2011]), and (c) visualization of data required for cut-off wall construction (Rosen et al. [2011]). GIS have been the primary data visualization medium in the dam safety studies and used to integrate various geo-databases to enable information exchange between such systems (Shumilov and Breunig [2000]). Apart from 2D- 3D visualization of behaviour of dam and site features and characteristics over time, engineers also prefer to easily access past construction photos and reports, in order to understand what features of the dam have changed over different phases of its life cycle. A very good example is the newly renovated Wolf Creek Dam in the US, where the instrumentation and construction data have been evaluated in relation to a geo-database and a 3D model of the dam. This integrated model supports engineers to analyse and predict the possible consequences of seepage and stability problems (Spencer et al. [2010]).

Several researchers have also worked on fully automated systems and web-based visualization techniques to facilitate quick feedback and information dissemination during multi-disciplinary meetings with participants from disparate locations (Lemke et al. [2011]). Continuously collected geotechnical, hydraulic and historical data assist engineers to build decision support system to analyse, supervise and forecast risk. For example, a web-based decision support system, which enables visualization of risk levels with colours, has been implemented to support estimating the flood failure risk of dikes (Maccabiani and Knoeff [2008]).

Although the previous studies are helpful to visualize and query information about dams in general and ease the decision making process, there is a lack of research studies in the domain that look at the dam risk assessment process and focus on developing an integrated shared knowledge repository for risk assessors. These studies did not focus on developing a holistic understanding of the ways engineers would like to look at the data, given their engineering discipline, and developing visual forms to enable those. The study presented in this paper focuses on characterization of such visualization needs to better serve engineers during their decision making processes while assessing risk levels of dams.

METHODS

The main objective of this study is to understand the requirements of the engineers with regard to their preferences in visualizing information while performing embankment dam risk assessment activities for a dominant failure mode. This paper provides findings in relation to internal erosion. Internal erosion, in particular, is complex to understand, and can even be triggered by normal day-to-day operations without a high intensity event like frequent high pool elevations. Internal erosion is also a major cause of failure of embankment dams (Blackett [2013]), and hence was the reason to focus on internal erosion in this study. Previous literature on requirements elicitation (Wiegers [1999]); Gould and Lewis ([1985]) suggests that the most productive approach to accumulate and analyze requirements for a specific task is to determine use cases and build prototypes with varying levels of details while utilizing user feedback at each stage of the prototype development process. The research team used a similar approach that incorporated a multi-phased requirement elicitation and case analysis to interact with engineers and document their visualization requirements during risk assessment process.

A three phased approach was used in this study to identify and validate the visualization requirements of engineers drawn from different disciplines. These phases are described in details in subsequent sections.

Phase 1: Requirements Elicitation through Systems Investigation and Interviews

In this phase, the research team conducted face-to-face unstructured interviews with engineers involved in risk assessment processes, and investigated the information systems used by the engineers to understand different views/figures currently generated with these systems. The larger goal of this phase is to compile a preliminary list of visualization requirements which would constitute an initial list of use-cases for a more-structured elicitation and validation of requirements. 15 engineers from different disciplines, as detailed in Table 2, participated in this study. Majority of these engineers were experienced engineers who have been involved in risk assessment processes for several embankment dams. Several systems are currently used by engineers to store, access, and visualize the collected sensor data. They gave integrated plotting, reporting and GIS-linking capabilities, based on predetermined templates. Various computational tools are used during risk assessments, including geographic information systems, geotechnical integration systems, 3D modelling systems and data viewers. During the study, these systems have been evaluated as part of the preliminary analysis so that the preliminary list of visualization requirements could be enumerated and that they could be communicated and discussed during the Phase I interviews with the engineers.

Table 2: Overview of participants of the study

Phase	Number of participants	Years of experience	Mean and standard deviation	Discipline (s)*
I	7	10-32	$\mu = 19.1$; $= 9.6$	H&H, GT, SE, GE, CE
II	5	13-37	$\mu = 27.6$; $= 9.7$	GT, H&H, GE, CNSTR, SE
III	2	4-16	$\mu = 10.0$; $= 8.5$	H&H, CE

*H&H: Hydraulic Engineering and Hydrology; SE: Structural Engineering; GE: Geology; GT: Geotechnical Engineering; CE: Civil Engineering; CNSTR: Construction Engineering.

Kasireddy et al.

Kasireddy et al. Visualization in Engineering 2015 3:1, doi:10.1186/s40327-014-0014-y

The primary focus of the interviews during Phase 1 was to capture discipline specific visualization requirements without delving too much into the process of extracting only those requirements which are relevant to the particular failure mode being assessed in this study. These preliminary findings were also useful to determine how engineers would like to visualize different dam features, and also to remove the ambiguity, if any, in the meaning of the terms from the perspectives of each engineering discipline.

Phase 2: Requirements Elicitation through A Card Game, Examination of Standards/ Guidelines and Case Documentation

Unlike the previous phase, wherein the requirements were collected in a generic sense, in this phase, the focus was particularly on assessment of internal erosion problems. In this regard, a card game was designed to expand the initial findings of the Phase I. Additionally, the team investigated standards and publications related to internal erosion assessment; and other risk assessment documentation available for three selected dams. The main strategy here is to corroborate the visualization requirements based on the analysis of multiple sources of information, i.e., through triangulation. Triangulation ensures the generality of the findings.

To approach capturing the discipline specific visualization requirements of engineers, a card game was designed to be used with accompanying scenarios. The card game included pile-of cards, and each card represented an information item that an engineer might be interested in knowing to understand the behaviour of a dam. Piles included several categories such as information about instrumentation, embankment features, historic reports, field tests, and drawings. Among each pile of cards, blank note cards were placed to accommodate the situation in which a participant asked for information that was not already represented in the pile of cards. A scenario represented a risk assessment case in a given dam setting where all the requested information by the participants assumed to be available. For every card (i.e., an information item) asked by a participant, participants were asked to define how they would like to visualize that information.

As part of the triangulation efforts, the research team examined various engineering guidelines/manuals like engineering manuals (EM), engineering regulations (ER). In addition, for three selected embankment dams, the research team examined the plots and visualization approaches used to depict or highlight identified facts about the dams in previous risk assessment reports.

Phase 3: Requirements Validation through Prototype Development and Face Validation

The main tasks carried out in this phase to validate the requirements identified in the above two phases included development of a functional prototype integrating all visualization requirements, and taking user feedback regularly through showing each identified and implemented view. The prototype was developed using an object oriented language and enabling renderings of rich 2D-3D graphics. With this prototype, it was possible to do face validation with the users in terms of pinpointing any discrepancies between what the research team interpreted versus what the users actually asked for.

For the face validation step, engineers provided their feedback on the identified visualization requirements to define whether the captured requirements represented what they described earlier and if they have any additional requirements to include. Regarding the feedback on the functional prototype, only two engineers participated on the evaluation a weekly basis for six months. The authors intend to perform functional evaluation of the prototype using a larger pool of engineers as the next step when they have access to a real Periodic Assessment and data about the dam being assessed.

RESULTS AND DISCUSSION

The findings are presented in terms of what has been identified as visualization requirements through the requirements elicitation approaches and then how the findings were implemented in the functional prototype.

Identified Visualization Requirements

The research team identified a total of 42 unique visualization requirements based on the research methodology outlined in the previous section. They have been tabulated in based on the engineering disciplines and the overarching categories of visualization. reveal that some of these discipline-specific requirements overlap with those of other disciplines, and the details of the same are discussed in the subsequent paragraphs. For the convenience of the reader, the authors have highlighted the overlapping requirements across different disciplines in bold font gives an idea of how visualization requirements vary with engineers from different backgrounds for the case of internal erosion risk assessment.

The distribution of the findings with respect to the engineering disciplines is not equally distributed. We can clearly understand from that 76% of the total unique visualization requirements were provided by geotechnical engineers and 38% of the requirements were provided by geologists, with overlapping requirements between groups. They were followed by Hydraulic engineers/Hydrologists (H&H group), who contributed to 17% of the total requirements. Similar in scale to the H&H group, structural engineers contributed only 14% of the total. The reason for having a wider set of requirements stated by geotechnical engineers and geologists is due to the scope of the problem being internal erosion, which falls more to the domain of geotechnical engineers. Also, since the scope of this study was limited to embankment dams in which structural features are minimal in comparison to other dam types such as the concrete dams, having a less number of requirements defined by structural engineers is expected.

When is analysed in terms of commonalities of visualization requirements based on engineering disciplines, it was observed that only 7% of the total 42 requirements such as geometrical information about dams; pre-existing structures; and reservoir pool and tail water elevations; were of interest to the engineers to look at collectively from all disciplines. There was a consensus among engineers regardless of their disciplines regarding certain visualization requirements. For example, all engineers preferred to have site plans for pre-existing features, which are important to know about for internal erosion assessment, around the dam site in 2D views. Similarly, the opinion was unanimous as far as the representation of dam geometry and information related to it in a 3D view. They also would like to have additional tools to be able to export different cross sections and plan views, and to turn on and off different layers (e.g., instrumentation, zoning, soil layers, pre-existing site plan, etc.). All disciplines also underscored the importance of visualizing the zoning within the dam (e.g., cross-hatching, colour, etc.) as well as the reservoir and tail water information. Here, all the engineers prefer to access the raw reservoir pool and tail water elevations and look at the related plots in a single view. In the same context, engineers would also like to be able to visualize water levels and flows over time (i.e., a 4D simulation of the water level on 3D dam geometry). In addition to these, the research team studied and identified that some of the requirements i.e., instrumentation information and readings provided within 3D settings and geotechnical and geologic information provided in plan views were common to at least three engineering disciplines.

Though there are overlaps in the visualization requirements among engineer disciplines, the percentage of overlap varies with the discipline specific visualization requirements. For instance, from it is evident that most of the 3D visualization requirements of geotechnical engineers overlapped with the requirements of the engineers from other disciplines. The overlapped features include turning on/off various layers of the information on the 3D model as well as visualization of instrumentation information (e.g., location, tip elevation, etc.) and instrumentation readings within the 3D settings. In contrast to that, the requirements of geologists do not have many overlaps with engineers from other disciplines.

Specific to the H&H group, hydraulic engineers were interested in the features enabling the visualization of regional rainfall inundation map, Possible Maximum Flood regional map, Hydro Meteorological Report-51 i.e. a probable maximum precipitation document, and the 3D view of the dam geometry. Furthermore, they also expressed interest in accessing tail water, pool elevation and reservoir inflow characteristics in a tabular form. Besides that, they also wanted to look at the hydrologic loading data for coincident pools for seismic PMFs, hydrologic loading data for flood events, inflow-volume-duration-frequency curve [1–7] day computed probability, pool-frequency, and pool-duration curves.

Incidentally, the visualization requirements of the structural engineers have a good overlap with those of the H&H group as far as the H&H tabular data is concerned. They have additional requirements for 3D visualization of the dam instrumentation and the site plan. On the other hand, the interests of civil engineers lie in the availability of instrumentation data - in the form of tables, and 3D geometry of the dam.

Implementation of Visualization Requirements in the Functional Prototype

The prototype was developed in an iterative and a participative manner, in which the opinion and feedback of the end users regarding the functionalities incorporated in the prototype, visual requirements implemented, and usability aspects, were regularly taken to customize existing features and also add new features if necessary. Initially, a view for accessing and displaying instrumentation meta-data was implemented along with a 2D data viewer for static 2D plots. A 3D model viewer was built in to the model and integrated with several required data to display contextual information about dam features and instrumentation data were added (Figure 5). In the next phase, querying capabilities for instrumentation data were incorporated. 2D data viewer was augmented with a dynamic time slider to visualize variation of readings over time, based on the feedback of engineers. In the following phases, views for bore-hole test results, document/photo access panels and image display capabilities were added to the prototype.

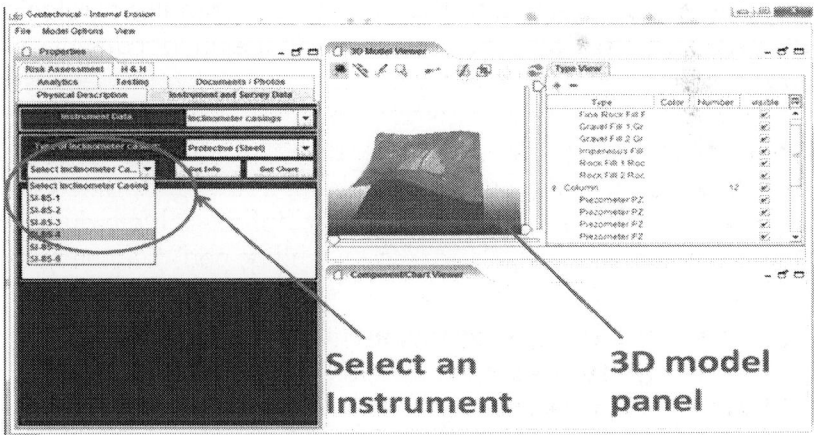

Figure 5: A snapshot showing 3D model panel.

Discussing all the features implemented in the prototype is out of scope of this publication; simply due to their sheer number and the space restrictions. However, some of them are detailed below.

Implementation for Visualization of Piezometer Meta-Data and Time-Series Readings

In relation to instrumentation data visualization, piezometers were the commonly referred instrument type to know about for internal erosion assessment. Engineers wanted to select different piezometric zones of influence within the 3D dam body and select the desired piezometers within them to examine their meta information. Meta-data and additional information to be specified for each piezometer included tabular and plotted piezometer data over time with respect to pool elevations, instrument location and tip elevation with respect to soil layers and stations in the dam, as well as piezometer influence zone in 3D phreatic surface. In addition to this, engineers would like to compare different piezometers using the querying functionality and plotting their readings over time along with the pool elevation variation using the time slider; and in the form of time series data were implemented– as shown in Figure 6. In the initial implementation, only the time series corresponding to the piezometers were plotted. However, the engineers indicated that it would be useful to them to understand any anomalies in piezometers if their readings were plotted alongside the reservoir pool levels. Thereafter, this feature was augmented to show even these details, on an as-desired basis.

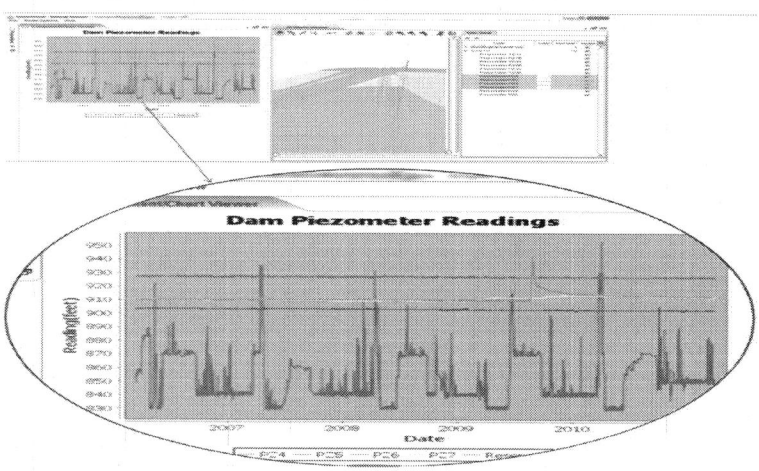

Figure 6: A snapshot showing a time series of selected piezometers.

Implementation for Visualization of Testing Data Such As Boring Logs and Rock Tests

Within testing data, "boring logs" is one of the frequently used words in the interviews with most of the geotechnical engineers and geologists. Important features implemented, concerning boring log information, are meta-data display of any selected bore hole inside a data panel; and display of different soil strata within each boring log. As engineers also showed tremendous interest in the ability to query for different bore holes based on a certain criteria, advanced query docking frame has been implemented for customized comparison, and here, users are able to put different bore-holes side-by-side and view their strata properties, and meta information and other related information (as shown in Figure 7). During validation of the functional prototype, after examining the implementation, the engineers mentioned their preference for the texture mapping of different soil strata with the conventional plot legends that they currently use in existing documentation. They believed such a feature would help them correlate strata, its properties and behaviour quickly, as they are already used to these conventions during their daily work routines.

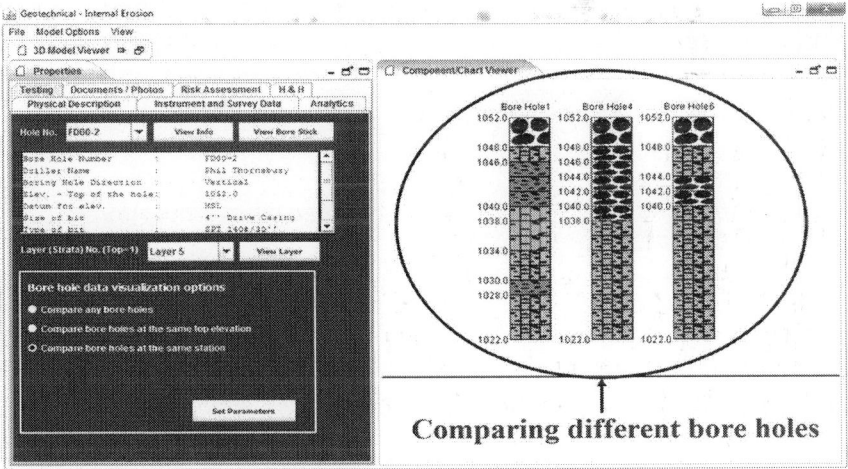

Figure 7: A snapshot showing that different bore holes can be compared (we can see different strata layers of each bore hole in this figure).

Implementation for Visualization of Documentation and Construction History Photos

Most of the dams have been constructed many years ago and they have lot of paper documentation concerning its construction history, repairs, site instrumentation, standards, etc. With time, it becomes very difficult to retrieve particular old documents, say, if needed for a risk assessment process or even for the perusal of the project engineers. Hence, engineers wanted an internal document indexing system within the prototype to drag and drop digitized files and photos and to be able to retrieve these indexed files quickly within the same interface, whenever needed. The implementation of this feature is shown in Figure 8, wherein a user selected a photo from the file index panel, and it is being displayed in the adjacent docking panel.

Figure 8: A snapshot is showing various documents and photos can be stored and accessed from the integrated prototype.

CONCLUSIONS

Visualization empowers engineers to conveniently visualize, integrate and accurately interpret the data from disparate sources.

For internal erosion risk assessment in embankment dams, engineers from several disciplines require dam information to be viewed from different perspectives. This study provides the findings of visualization requirements of engineers involved in risk assessment processes while looking at historical dam information.

While the engineers would like to be able to use the current methodologies they are using to visualize static data related to embankment dams, they desire for an advanced 3D visualization paradigm that allows the end users to at least import different cross sections and plan views; turn on and off different information layers concerning instrumentation and other site plans; and simultaneous comparison through querying and visualization of multiple boring logs, piezometers, and monuments.

The findings from this study suggest that engineers would like to visualize the dam layout, components and geometry within 3D settings overlaid with sensor data, and querying capabilities in order to get a better flexibility to understand the risk associated with potential failure modes. The authors believe that, armed with this flexibility, engineers can be more effective and efficient during risk assessment sessions, and can contribute to better dam maintenance decisions. However, the validation of effectiveness and efficiency for decision making is a required next step. Validation will be more effective when performed during an actual periodic assessment of a dam. The main contributions of this paper comprise the visualization requirements in the domain of dams, which has not been attempted before. However, the same research methodology can be utilized to extract requirements in other decision domains.

Future work can include putting efforts to quantify the value of using visualization tools, discussed in the study with engineers, through scenarios from a specific dam for assessment of internal erosion during a real PA exercise. Among the instruments mainly used in the data collection tasks at the dam location, the research team focused mainly on the piezometers in the risk assessment process for the current study. In the future, other available instrumentation and their readings could be investigated to understand internal erosion risk and risk due to other failure modes in a holistic manner.

AUTHORS' CONTRIBUTIONS

VK contributed to conception and design; analysis and interpretation; data collection; critical revision of the article; statistical analysis; and takes the overall responsibility of the paper. SE participated in conception and design; analysis and interpretation; data collection; writing the article; critical revision of the article; statistical analysis; and obtained funding for the project. BA participated in conception and design; critical revision of the article; and obtained funding for the project. NSG participated in analysis and interpretation; data collection; writing the article; critical revision of the article; and statistical analysis. All authors read and approved the final manuscript.

ACKNOWLEDGEMENTS

The research presented in this paper is supported by USACE grant. The authors would like to acknowledge the support of Chris Kelly, Meghann Wygonik and other engineers from USACE who participated in this research at various stages. Also, we developed the prototype using the research version of java toolbox provided by IFC Tools Project (Tulke et al. [2013]).

REFERENCES

1. American Society of Civil Engineers (2013). America's Infrastructure Report card. On-line: http://www.infrastructurereportcard.org. Accessed: 27 Jan 2014.

2. Blackett, F (2013). Potential Failure Modes for Piping and Internal Erosion. On-line: http://www.oregon.gov/owrd/SW/docs/dam_safety/M6_Blackett_Internal_Erosion.pdf. Accessed: 29 Jan 2014.

3. Brindley TL, Tarantino JJ, Locke AL, Dollins DW: Utilization of 4-Dimensional Data Visualization Modeling to Evaluate Burial Ground Contaminants at the Paducah Gaseous Diffusion Plant, Paducah, Kentucky. In *Waste Management 2006 Symposium*. WM Symposia, Inc, Tucson; 2006.

4. Cross, B, Rogoff, E, Fricke, L (2005). Data Visualization Tools for Litigation - Practical Uses and Ethical Considerations. Proceedings

of the 2005 NGWA Ground Water and Environmental Law Conference, Baltimore, MD, 1–14

5.   Dominguez-Acosta, M, Granados-Olivas, A, Hibbs, B, Eastoe, C, Hawley, J (2004). Computer Based Three-Dimensional Modeling of Hydrogeologic units in the Transboundary Ciudad Juárez-Paso del Norte region. Second International Symposium on Transboundary Water Management Proceedings. Tucson, AZ

6.   Glynn, P, Jacobsen, L, Phelps, G, Bawden, G, Grauch, V, Orndorff, R, Winston, R, Fienen, M, Cross, V, Bratton, J (2011). 3D/4D Modeling, Visualization and Information Frameworks: Current US Geological Survey Practice and Needs. In Three-dimensional Geologic Mapping Workshop, Minneapolis, MN, 33–39

7.   Gould J, Lewis C: Designing for Usability: Key Principles and What Designers Think. *Comm of the ACM* 1985, 28(3):300-311.

8.   Lai JS, Chang WY, Chan YC, Kang SC, Tan YC: Development of a 3D virtual environment for improving public participation: Case study–The Yuansantze Flood Diversion Works Project. *Advanced Engineering Informatics* 2011, 25(2):208-223.

9.   Lemke J, Driller M, Wilson D: Web-based Real-time Monitoring at Perris Dam Using In-place Inclinometers and Piezometers with an Automatic Notification System. In *21st Century Dam Design - Advances and Adaptations, 31st Annual USSD Conference.* United States Society on Dams, San Diego, CA; 2011:1527-1540.

10.   Maccabiani J, Knoeff JG: An online tool for real-time analysis and management of flood risks of diked areas. In *4th Int. Symp. Flood Defence.* International Flood Initiative, Toronto, Canada; 2008:1-8.

11.   Pantea MP, Hudson MR, Grauch VJS, Minor SA: *Three-Dimensional Geologic Model of the Southeastern Española Basin, Santa Fe County, New Mexico. U.S. Geological Survey Scientific Investigations Report 2011–5025.* USGS Geology and Environmental Change Science Center, Denver, CO; 2011.

12.   Qi H, Altinakar MS: A GIS-based decision support system for integrated flood management under uncertainty with two dimensional numerical simulations. *Environmental Modelling & Software* 2011, 26(6):817-821.

13. Rosen, JB, Arnold, MA, Bachus, RC, Schauer, D, & Berrios, A (2011). GIS for Geotechnical Decision Making: Visualization of Cut-Off Wall Construction Data. In Geo-Frontiers 201: Advances in Geotechnical Engineering (pp. 2907–2916). ASCE.

14. Serre D, Peyras L, Tourment R, Diab Y: Levee performance assessment methods integrated in a GIS to support planning maintenance actions. *Journal of Infrastructure Systems* 2008, 14(3):201-213.

15. Shaffner, P (2011). Geologic Data and Risk Assessment; Improving Geologic Thinking and Products. 21st Century Dam Design – Advances and Adaptations, 31st Annual USSD Conference, San Diego, CA, 545–569

16. Shumilov S, Breunig M: Integration of 3D geoscientific visualization tools with help of a geo-database kerne. In *Proceedings of the Sixth EC-GI & GIS Workshop*. The Spatial Information Society–Shaping the Future, Lyon, France; 2000:66-76.

17. Spencer, W D, Fritze, B, Greene, D C, and Haskins, T A (2010). The Use of Electronic Data Anaylsis and 3-D Modeling to Make Us Smarter. 30th Annual USSD Conference, Sacramento, CA, 653–681

18. United States Army Corps of Engineers: *Improving Safety Monitoring for Embankment Dams*. CMU, Pittsburgh, PA, USA; 2012.

19. Wiegers K: First things first: prioritizing requirements. In *Software Requirements*. 2nd edition. Microsoft Press, Redmond, WA; 1999.

20. Xi-juan J, Rui-sheng J, Jing-bo Z: Multi-layer DEM modeling and application in stratum visualization based on IDL. In *Computer Application and System Modeling (ICCASM), 2010 International Conference*. IEEE, Taiyuan, China; 2010:150-154.

21. Tulke, J, Tauscher, E, Theiler, M, and Thomas, R. Open IFC Toolbox. On-line: http://www.ifctoolsproject.com. Accessed: 30 Aug 2013

Characterization of Ambient Seismic Noise near A Deep Geothermal Reservoir and Implications for Interferometric Methods: a Case Study in Northern Alsace, France

Maximilien Lehujeur, Jérôme Vergne, Jean Schmitt-buhl, and Alessia Maggi

IPGS, Université de Strasbourg/EOST, CNRS, 5 rue René Descartes, Strasbourg Cedex, 67084, France

ABSTRACT

Background

Ambient noise correlation techniques are of growing interest for imaging and monitoring deep geothermal reservoirs. They are simple to implement and can be performed continuously to follow the evolution of the reservoir at low cost. However, these methods rely on assumptions of spatial and temporal uniformity of seismic noise sources. Violating them can result in misinterpretation of seismic velocities owing to preferential noise propagation directions.

Methods

Using several years of seismic data recorded around the two geothermal sites of Soultz-sous-forêts and Rittershoffen in northern Alsace, France, we propose a detailed characterization of the spatial and temporal properties of the high frequency seismic noise (0.2 to 5Hz). We consider two fundamental properties of the cross correlation functions (CCFs) of ambient noise. Firstly, the reliability of the Green's function reconstruction, an important indicator for tomographic studies. Secondly, the temporal repeatability of the CCFs between 0.2 and 0.5 seconds.

Results and Conclusions

At periods below 1s, we observe a sharp decrease in signal to noise ratio resulting from the non-uniform distribution of anthropogenic sources. At periods above 1s, we show that the high directivity of the northern Atlantic microseismic peak biases the CCFs' phase significantly. We show that nocturnal noise is the most suited for temporal analysis of the CCFs. Using nocturnal noise, we should be able to monitor temporal variations induced by the geothermal activities inside the reservoir.

BACKGROUND

Projects dedicated to the exploitation of deep geothermal resources need to probe the upper crustal structure of the targeted area in order to characterize the reservoir and its relation to pre-existing geological formations. Active seismic sounding is a commonly used approach; its dense spatial and temporal sampling provides high-resolution images of the reflectivity of the subsurface layers and of fault geometry. However, such seismic data are not readily available everywhere, and acquisition of new data, especially in 3D, is often very expensive compared to the profitability of geothermal resources. Its high cost also excludes using repeated active seismic sounding to follow the evolution of the medium during production.

Images of the upper crustal structure can also be obtained from tomographic inversion of arrival times of natural or induced local earthquakes. These inversions can be repeated over time to map velocity changes (e.g., Calò et al. [2011]; Calò and Dorbath [2013]) much more cheaply than active seismic sounding. Producing good-quality tomographic images of geothermal reservoirs using arrival times requires having induced seismicity around the wells, but geothermal operators need to minimize induced seismicity to reduce the seismic risk associated with their exploitation. The need to minimize seismic risk excludes using time-lapse arrival-time tomography for continuous reservoir imaging.

Over the past 10 years, another promising passive seismic imaging technique has emerged. Known as 'ambient noise tomography', it uses seismic noise as a permanent source of energy that propagates through the target region. The cross-correlation function (CCF) of long records of seismic noise at a pair of stations provides an estimation of Green's function between them (Lobkis and Weaver [2001]; Shapiro and Campillo [2004]; Sabra et al. [2005a]). The resulting correlogram is similar to the signal that would be obtained if an impulsive source occurred at one station and was recorded by the other one. This method allows us to perform tomographic studies using all possible pairs of stations over a network (Shapiro et al. [2005]; Sabra et al. [2005b]). The resolution of the recovered seismic velocity models only depends on the geometry of the stations. This technique has been widely applied at various scales, from the structure of the mantle using worldwide broadband stations (Poli et al. [2012]; Lin and Tsai [2013]; Lin et al. [2013]) to laboratory samples using piezoelectric sensors (e.g., Lobkis and Weaver [2001]; Derode et al.[2003a], [b]; Larose et al. [2007]). At the local scale, the method has been applied in various environments from offshore oil reservoirs (Bussat and Kugler [2011]; Mordret et al. [2013]) to active volcanic systems (Brenguier et al. [2008], [2011]). Although the applicability of ambient noise tomography in the context of geothermal reservoirs is still under debate, at least one application has already been attempted using a local network of short-period seismometers around the geothermal site of Soultz-sous-Forêts (Calò et al. [2013]). Beyond its use in seismic tomography, the continuous nature of seismic noise can also be exploited to observe subtle variations in the seismic velocity or the diffracting character of the crust. For example, Brenguier et al. ([2008], [2011]) and Obermann et al. ([2013]) were able to produce 4D pictures of the Piton de la Fournaise volcano by applying interferometric analysis to the coda part of the correlograms.

The seismic noise-based methods described above all rely on strong assumptions concerning the noise sources. For tomography applications, noise sources should be homogeneously distributed (Lobkis and Weaver [2001]; Roux et al. [2005]). Under this assumption, only the sources located in a narrow area along the continuation of the path joining the two stations contribute to the recovered Green's function (Roux et al. [2004]; Sabra et al. [2005c]; Larose [2005]; Gouédard et al. [2008]). For applications that monitor time-dependent perturbations of the medium, noise sources may be inhomogeneously distributed, but in this case, they must be repeatable. If the seismic noise sources move too much over time, the resulting changes in the signal could be mistaken for perturbations of the medium (Hadziioannou et al. [2009]; Weaver et al. [2011]).

The consequences of violating these assumptions have been studied theoretically and numerically using synthetic data (Weaver et al. [2009]; Froment et al. [2010]). Although the seismic noise distribution on Earth is rarely homogeneous in time and space, the CCFs approximate Green's functions correctly if the inter-station distances are long and the azimuthal distribution of the noise is smooth. However, when the noise source distribution is highly heterogeneous, some studies using real data report significant bias and incorrect estimation of seismic velocities between the station pairs (Pedersen and Krüger [2007]).

Here, we focus on the application of the ambient noise correlation technique in the context of a geothermal reservoir (i.e., a kilometer scale) using seismic data around geothermal sites in northern Alsace. We analyze the characteristics of the seismic noise in the period range between 0.2 and 5 s (0.2 to 5 Hz). We then study the correlograms of ambient noise records between pairs of stations and show how the seismic noise distribution influences the quality and reliability of the reconstructed Green's functions in this particular period range. Finally, we examine the stability of the high-frequency CCF coda for future analysis of temporal changes within the reservoir.

DATA

The Upper Rhine Graben concentrates several sites dedicated to the exploration and exploitation of deep geothermal energy. Northern Alsace hosts both the prototype site of Soultz-sous-Forêts which was initiated about 20 years ago and a recent industrial project in Rittershoffen (ECOGI), 10 km to the southeast, which started in 2012 and is expected to reach its exploitation phase in 2015. Seismometers have been deployed around both sites to monitor the natural and induced seismicity (Figure 1). They form a permanent network of 12 short-period stations equipped with 1-Hz L4C sensors and digitizers sampling at rates from 100 to 200 Hz. In this study, we use the high-quality continuous recordings available since 2010 for the Soultz-sous-Forêts network and since 2012 for the Rittershoffen network. We also include data from a temporary network of 16 short-period sensors (1-Hz corner frequency) installed in May 2013 by the Karlsruhe Institute of Technology and Ecole et Observatoire des Sciences de la Terre of Strasbourg to densify the permanent network around the site of Rittershoffen (purple network, Figure 1; Gaucher et al.[2013]) during stimulation of the GRT1 well. Taken altogether, these stations form a 15-km-wide seismic array, whose station spacing ranges from 1 to 15 km.

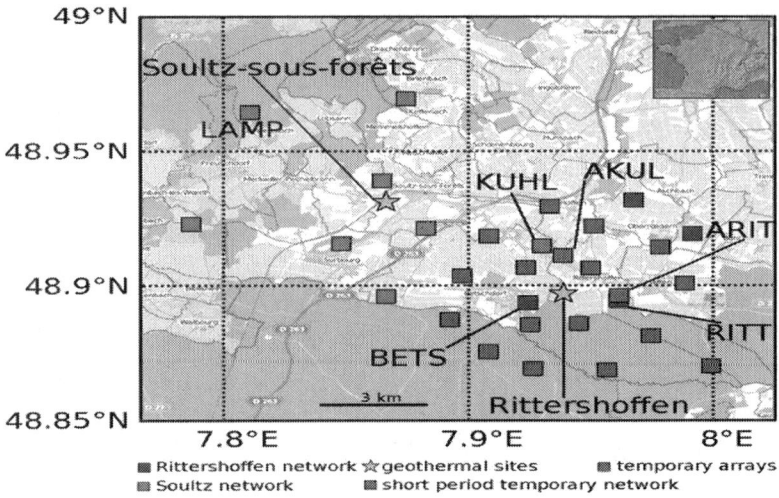

Figure 1: Map of geothermal sites (Soultz-sous-Forêts and Rittershoffen) and available seismic stations in the area. ©OpenStreetMap contributors.

In order to extend our understanding of the origins of seismic noise to higher frequencies, we deployed two small aperture arrays, ARIT and AKUL, close to the location of the permanent stations RITT and KULH (Figure 1). The arrays operated for 2 months during fall 2012. Each array contained six vertical short-period sensors (1-Hz corner frequency) with one three-component L4C sensor at the center. All the sensors were connected by cables to a central acquisition system that provided a common time reference for the nine recorded channels. They were deployed in a helical configuration with a 300-m maximum aperture.

METHODS

Frequency Content and Temporal Variability of Seismic Noise

Empirical Green's functions constructed from the correlation of vertical component ambient seismic noise records are dominated by surface waves, because most noise sources occur close to the Earth's surface.

Therefore, the dispersive character of the Rayleigh waves is the primary information that can be extracted from correlograms (Campillo et al. [2011]). Dispersion measurements for each pair of stations (group or phase dispersion curves) can be regionalized to provide spatial variations of surface-wave velocities at each period. Inversion can then be used to convert surface-wave velocities as a function of period to S-wave velocities as a function of depth (the longer the surface-wave period, the greater the investigation depth). At Soultz-sous-Forêts and Rittershoffen, the reservoir lies between 2- and 5-km depths. In order to map S-wave velocities from the surface to that depth using the dispersive character of the Rayleigh waves, we must work in the period range of 0.2 to 5 s. This range is compatible with the bandwidth of our seismometers (cutoff period of 1 s) and benefits from low instrumental noise.

Between 0.2 and 5 s, seismic noise has differing origins and properties. For periods above 2 s, seismic noise spectra everywhere on Earth contain a broad, highly energetic peak called the 'secondary micro-seismic peak.' This peak results from pressure variations on the sea bottom induced by interferences of oceanic waves traveling in opposite directions (Longuet-Higgins [1950]). A few dominant zones in the north Atlantic (south of Greenland, along the Canadian coasts and around the mid-Atlantic ridge) generate most of the secondary micro-seismic peak energy recorded in Europe (Gutenberg [1936]; Kedar et al. [2008]; Sergeant et al. [2013]).

At periods below 2 s, numerous phenomena are responsible for the observed seismic noise. They can be split into two categories: natural sources, among which the wind acting on trees or structures sealed into the ground in the vicinity of the recording stations (Withers et al. [1996]; Bonnefoy-Claudet et al. [2006]), and anthropogenic sources like road traffic, industries, or other types of human activities (McNamara and Buland [2004]; Groos and Ritter [2009]). As seismic noise from these high-frequency sources propagates only to local distances, its characteristics change from one region to another. A region-specific analysis of the high-frequency noise spectrum is therefore recommended before applying ambient noise correlation techniques (Campillo et al. [2011]).

Figure 2 presents spectrograms of the noise recorded at station RITT. The columns in each spectrogram represent the amplitude of the

Fourier transform of the ground acceleration computed for 1 h of signal. The 1- to 10-s period range (above the white dotted line) is dominated by the secondary micro-seismic peak; its maximum amplitude occurs between 2 and 7 s. The amplitude of this noise is independent of the recording site, proving that it is produced by distant sources and is recorded coherently over the whole network. We observe an annual periodicity, with an increase of energy during the winter (white arrows labeled W). Below 1 s, the noise is very different and presents a strong daily periodicity with a decrease of energy during the night (white arrows labeled N) and at noon (white arrows labeled 12). We also observe a weekly periodicity with significant noise reduction during the weekend (labels 'Sat' and 'Sun'). These periodicities underline the dominant role played by anthropogenic sources in this period band. Even though the same overall features are observed on most stations, the energy and detailed characteristics are highly dependent on the recording site, confirming that this high-frequency noise is generated close to the stations.

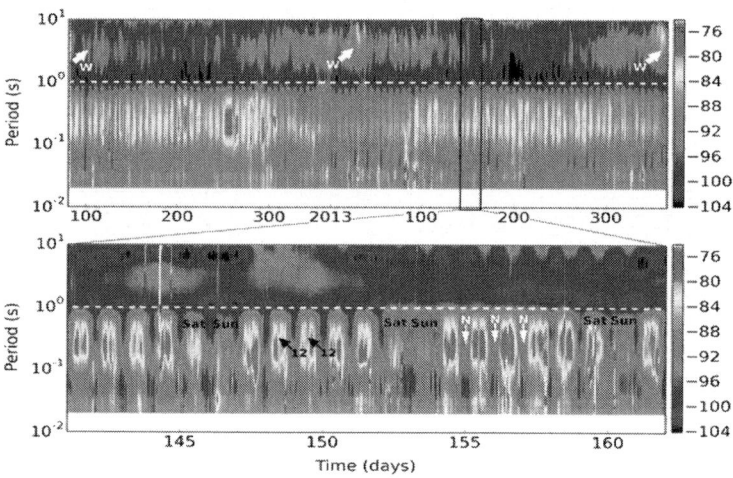

Figure 2: Vertical component spectrograms of over 2 years of data at station RITT. Top: full data set. Spectra are averaged per 24 h. Bottom: zoomed-in image showing nearly 3 weeks of data. The color corresponds to the modulus of the ground acceleration expressed in decibels (20 log (m.s^{-2}.Hz^{-1})). The white dotted line at 1-s period separates the noise dominated by the secondary micro-seismic peak (above the line) from the noise dominated by anthropogenic activity (below the line). White arrows labeled 'W' indicate stronger

seismic noise occurring during winter. Arrows labeled as 'N' and '12' indicate weaker anthropogenic noise occurring during the night and at noon, respectively. 'Sat' and 'Sun' labels stand for Saturday and Sunday, respectively.

Spatial Distribution of Seismic Noise

To estimate the spatial origin of the seismic noise over the whole period range of interest, we apply a classical beamforming technique (e.g., Rost and Thomas [2002]) to the local monitoring network and the small aperture arrays. This technique allows us to determine the dominant back azimuth and phase velocity of an incoming seismic wave, so long as the network's station spacing is less than half the wavelength. By applying the method on a sliding window, we can estimate the directivity and phase velocity of the noise as a function of time. The longer the period of the noise to be processed, the wider the array must be.

The permanent network of Soultz-sous-Forêts can be used as a single array for periods between 2 and 5 s. We estimate the most probable incoming direction and phase speed of the noise over 4 years of continuous records divided into 15-min windows. The results are given as a probability density function in the phase speed-back azimuth domain. Figure 3 shows an example for 2-s period. At this period, we observe a dominant phase velocity of approximately 3 km/s, corresponding to the average phase speed of Rayleigh waves under the Soultz network. Over 95% of the noise arrives from back azimuths between 265° and 345°, which correspond to the direction of the northern Atlantic Ocean. The other periods between 2 and 5 s yield similar results. If we group the beamforming results as a function of the month of the year, we observe a coherent annual variation of ±5°, which is probably related to small seasonal variations in the location of the noise sources in the Atlantic Ocean (Figure 3, bottom).

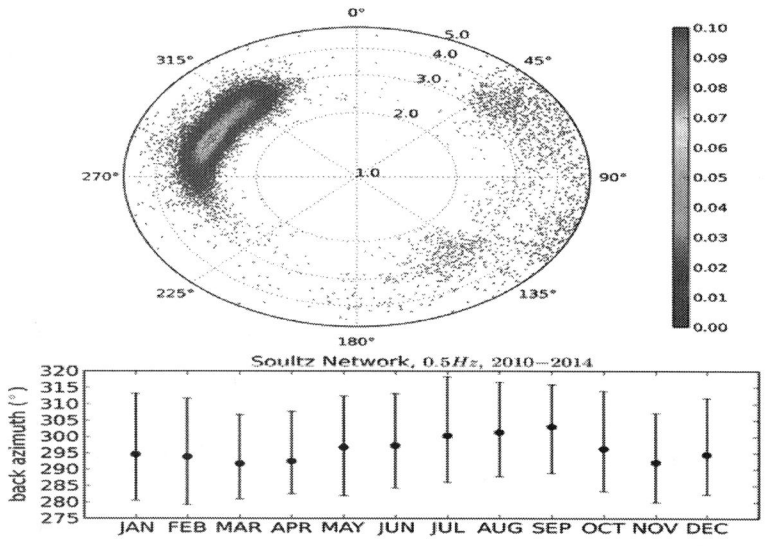

Figure 3: Beamforming analysis performed on the Soultz-sous-Forêts network. About 4 years of data have been divided up into 15-min windows and filtered around 1s. Top: probability density function (color code) of the estimated phase speed (expressed in km/s and displayed as the radius) and back azimuth (measured clockwise from the north in degrees) of the noise. Bottom: detail of the variation of the dominant back azimuth measured using the 4 years of data grouped by month (all January months together, etc.). Error bars delimitate the 16% and 84% percentiles of the statistical distribution of detected back azimuths.

Below 1 s, beamforming can no longer be performed on the full network because the station spacing is too large compared to the wavelength of the seismic noise, which results in aliasing effects. For this reason, we use the two small aperture arrays AKUL and ARIT to identify the origin of the noise at periods below 1 s. We estimate the most probable incoming direction and phase speed of the noise from 2 months of continuous recording divided up into 15-s windows. Figure 4presents the results of the beamforming analysis at a period of 0.3 s. The radial histograms of measured back azimuths are normalized to the total number of 15-s windows and superimposed on the map of the area at the array locations (left side of Figure 4). Note that these histograms do not provide any information about the energy of the seismic sources but only their statistical distribution in azimuth. The azimuthal distribution of seismic sources is heterogeneous with few narrow peaks, indicating that the high-frequency noise around the arrays is mostly generated in two zones that roughly correspond to the neighboring villages. Furthermore, the relative weight of these zones varies with the time of day (right side of Figure 4). For instance, the source detected by array AKUL at back azimuth 140° - the village of Rittershoffen - becomes prominent during the day. Other sources can only be observed at night, such as the one detected by array AKUL at back azimuth 50°, which does not point towards any particular village. These sources may be continuous in nature but are masked during the day by other dominant sources.

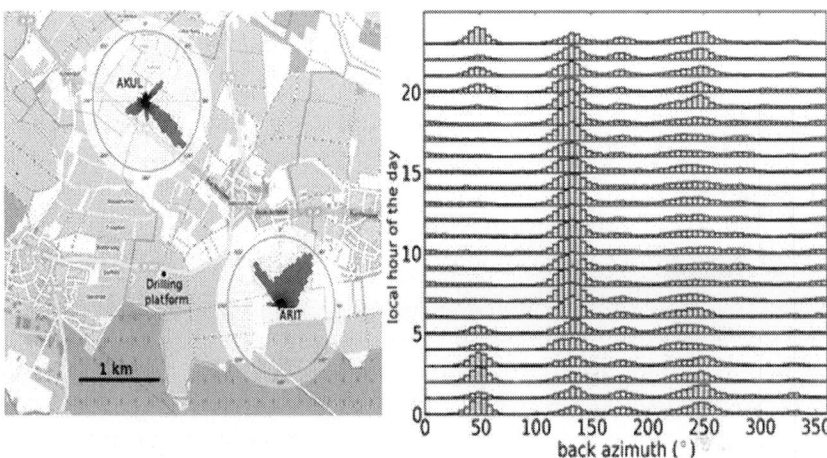

Figure 4: Beamforming analysis performed at the two small aperture arrays ARIT and AKUL at 0.3-s period (3.3 Hz). Left side: polar histograms showing the relative number of detected arrivals with back azimuth. The histograms are superimposed to the map of the area at the array locations, ©OpenStreetMap contributors. Right side: evolution of the histogram of detected arrivals with the hour of the day at array AKUL. All histograms are normalized to the total number of detected arrivals during each hour.

RESULTS AND DISCUSSION

Cross-Correlation Functions and Dispersion Measurements

We compute CCFs for each possible pair of stations of the full network (excluding the small aperture arrays). Each 1-h-long segment of noise is processed individually prior to correlation. Following the sign convention proposed by Stehly et al. ([2006]), we systematically correlate the noise recorded at the eastern station of a pair with the western one. Consequently, the noise sources occurring west (resp. east) to the two stations affect the positive (resp. negative) part of the CCF. We band-pass filter the CCFs in three period ranges and represent them according to the inter-station spacing (hodograms; Figure 5). Between 1.25 and 5 s (0.2 to 0.8 Hz; Figure 5, top left), the Rayleigh waves can be unambiguously identified on almost all station pairs. The stronger amplitudes of the positive parts of the correlations are a consequence of the directivity of the seismic noise (Stehly et al. [2006]). Indeed, as shown by the beamforming analysis, this noise mainly originates from the northern Atlantic Ocean located west-northwest of the network. Around 1 s (Figure 5, top right), the Rayleigh waves can still be followed on the hodogram. The increased symmetry indicates that, in this period range, noise sources are more uniformly distributed around the network or that the noise energy is efficiently distributed through the medium via scattering. However, the signal-to-noise ratio is significantly lower than that in the 1.25- to 5-s period range. Finally, at periods below 1 s (Figure 5, bottom left), no coherent wave field can be clearly observed, and the signal-to-noise ratio is very low.

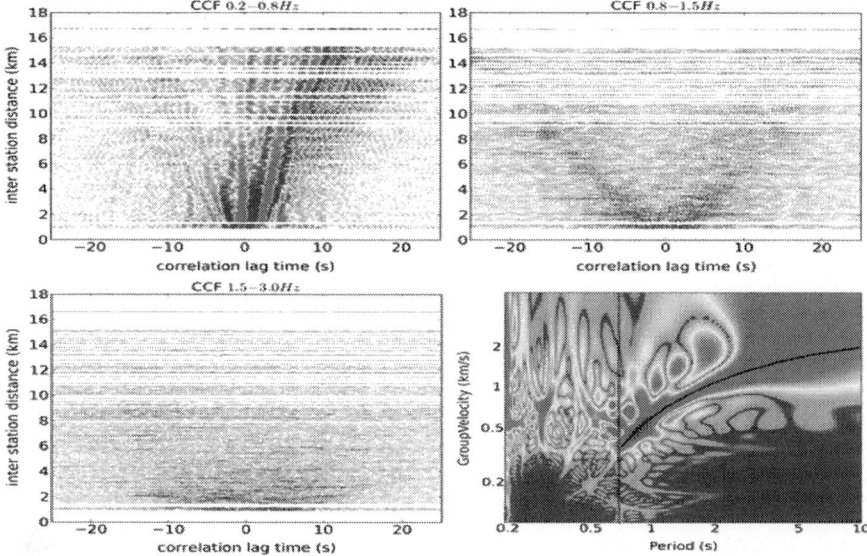

Figure 5: Band-pass filtered cross-correlation functions and example of dispersion diagram. The CCFs are computed for each possible station pair and band-pass filtered between 0.2 and 0.8 Hz (1.25 to 5 s, left), 0.8 and 1.5 Hz (0.66 to 1.25 s, right), and 1.5 and 3 Hz (0.33 to 0.66 s, bottom left). Bottom right: dispersion diagram obtained by frequency-time analysis of the KUHL-BETS station pair. The solid line indicates the fundamental mode of the Rayleigh waves. The color code corresponds to the energy of the signal in the group velocity versus period domain in arbitrary units (red corresponds to high energy and blue means no energy).

These observations are confirmed by the dispersion analysis performed on individual CCFs. We measure the Rayleigh wave dispersion on the CCFs by frequency time analysis (FTAN), which provides an estimation of the group velocity at each period (e.g., Dziewonski et al. [1969]; Bensen et al. [2007]). Figure 5 (bottom right) illustrates a typical dispersion diagram obtained for the station pair KUHL-BETS. Despite its noisy aspect, we can clearly identify the dispersion of the fundamental mode Rayleigh wave at periods longer than 1 s (solid black curve on Figure 5, bottom right). This mode can be easily identified on most station pairs. The dispersion diagrams change markedly around 1 s, and we cannot estimate group velocities at shorter periods. This inability occurs regardless of the chosen station pair and the time range (from weeks to years) used to compute the

CCF. This transition period of approximately 1 s is similar to the one observed on the seismic noise spectrograms (Figure 2) and corresponds to the transition between the noise originating from oceanic sources and that generated by local anthropogenic activities.

We propose two possible explanations for the poor quality of the reconstructed Green's function at periods below 1 s: (1) because of attenuation, the noise generated by low-energy local sources cannot travel far enough to be coherently recorded by two separate stations and/or (2) the non-uniform distribution of the local sources limits the reconstruction of Green's function, as theoretically predicted. The consequences of non-uniform noise sources are described in the following section.

Impact of a Directive Noise

We observe high signal-to-noise ratio on the phase of the CCFs between 0.2 and 1 s (Figure 5, top left). We should therefore be able to estimate phase velocities for each station pair and invert them to map their geographical variations. We start by estimating the average Rayleigh wave phase speeds over the whole network at each period by looking for phase alignments among band-pass-filtered CCFs in the time-distance domain (slant stack technique). Figure 6 (top left) presents the CCFs band-pass filtered around 2 s; the dashed lines correspond to the estimated average phase speed (2.45 km/s). We then measure the time shift between the phase arrival time picked on each CCF and the one predicted by the average dispersion law (time delay measured between the phase of each CCF and the dashed lines on Figure 6, top left). Negative time shifts correspond to phases arriving sooner than predicted by the reference phase velocity. In the ideal case, these time delays should only be caused by spatial variations of the phase velocities.

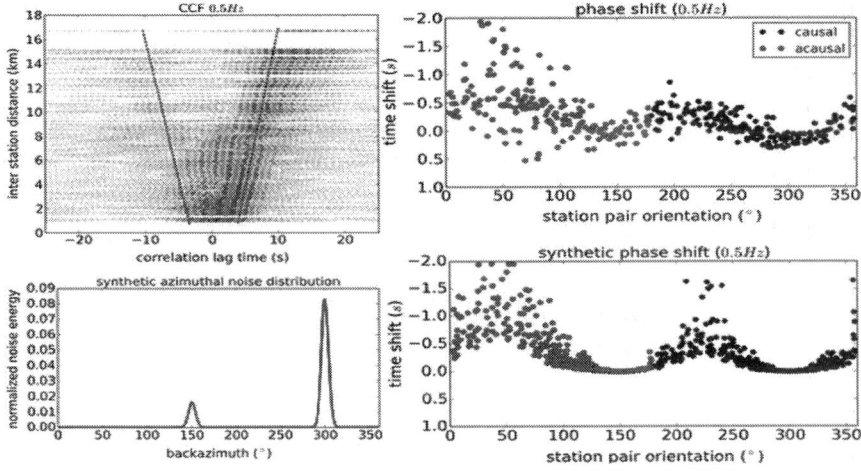

Figure 6: Impact of the noise directivity on the CCF phase. Top left: cross-correlation functions band-pass filtered at 2-s period (0.5 Hz). The dashed lines indicate the phase speed of the Rayleigh waves averaged over the full network and measured via slant stack (2.45 km/s). Top right: time shift measured between the phase arrival time of each CCF and the time predicted by the average phase speed. A negative time shift corresponds to a phase arriving before the predicted one. Because of the chosen orientation convention, phase shifts measured on the positive (resp. negative) part of the CCFs are attributed to azimuths ranging from 180° to 360° (resp. 0° to 180°). Bottom left: synthetic distribution of the noise energy with back azimuth. Amplitudes are normalized so that the area below the curve is 1. Bottom right: synthetic time shifts predicted using the formula from Weaver et al. ([2009]) at 2 s in a homogeneous medium (2.45 km/s).

We display the measured time shifts as a function of the orientation of the station pair (Figure 6, top right). Because of the chosen orientation convention, phase shifts measured on the positive (resp. negative) part of the CCFs are attributed to azimuths ranging from 180° to 360° (resp. 0° to 180°) and are caused by eastward (resp. westward) propagating noise. We observe a sinusoidal shape with a minimum occurring for station pairs oriented approximately 300°, corresponding to the case in which the station pairs are aligned with the dominant direction of oceanic noise in this period range (Figure 3). Even though these phase shifts might be due to spatial variations of the phase speed at depth, the sinusoidal-shaped variation combined with a minimum phase shift at 300° suggests that the phase shifts might in reality be caused by noise directivity. In another context, Pedersen and Krüger ([2007]) observed apparent variations of the group speed that were actually caused by strong noise directivity.

To corroborate the claim that our phase shifts are essentially caused by noise directivity, we propose a synthetic test with a homogeneous medium (constant phase speed of 2.45 km/s at 2-s period). Using the work of Weaver et al. ([2009]) and Froment et al. ([2010]), we estimate the theoretical phase shifts predicted by a given azimuthal distribution of the noise energy (ADNE). We find that a simple synthetic ADNE made of two Gaussian functions centered at 150° and 300° (the principal back azimuths of the noise directivity observed in Figure 3) reproduces the main features (sinusoidal shape and amplitudes) of the phase shifts measured on real data. Because the synthetic medium is homogeneous, these phase variations can be attributed unambiguously to the noise directivity. Interestingly, as soon as the ADNE contains an isotropic component, however small (even 0.01% of the dominant arrival), the phase shifts become negligible. This confirms the main conclusion of Weaver et al. ([2009]) and implies that a complete lack of coherent isotropic noise strengthens our observed phase variations.

The predicted time shifts in our synthetic test vary from 0% to +40% of the time needed by the phase to travel between two stations, depending on the distance separating the stations and how the pair is oriented with respect to the incoming noise. As our network is quite dense, each grid cell is observed by several station pairs having different orientations and separations. In a tomographic application, one can naively expect to restore the uniform distribution of the noise with azimuth by canceling out the noise-induced phase shift. To understand how the error on the phase of the CCF spreads over the spatial distribution of phase velocities through tomography, we invert the biased synthetic propagation times (Figure 7). We assume a homogeneous a priori velocity of 2.45 km/s, and set the smoothing distance to 5 km. As the computed phase shifts are all positive, most grid cells exhibit overestimated phase speeds that can reach 30% more than the true value (green/blue zones). These artefacts are heightened by the large number of station pairs oriented perpendicularly to the dominant noise arrival (300°). The inverse problem also produces zones of underestimated phase speeds (yellow zones) to balance the overestimations and satisfy most of the observations. Future work will focus on how to take this bias into account for a reliable estimation of velocity variations at depth.

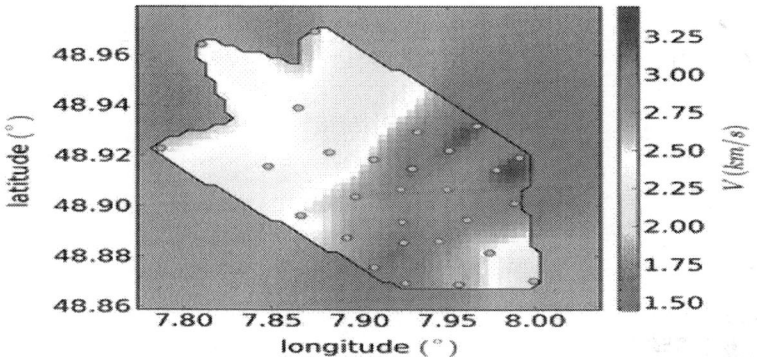

Figure 7: Tomographic inversion of the theoretical time shifts induced by the synthetic anisotropic ADNE. Yellow dots correspond to the stations used in the inversion. The ideal model is homogenous with a phase speed of 2.45 km/s (white color). Green/blue (resp. yellow) colors correspond to over-estimate (resp. under-estimated) phase speeds; both causal and acausal time shifts are included in the inverse problem. The black line delimits the resolved zone.

Impact of Localized and Repetitive Deterministic High-Frequency Sources

Since the energy of short-period noise (below 1 s) presents a strong weekly and daily periodicity resulting from human activity (Figure 2, bottom), we average the short-period CCFs (0.2 to 0.5 s) separately for each day of the week. Figure 8 illustrates the evolution of the CCFs for the 7 days of the week for the station pair RITT-BETS (2.8 km apart). We observe that the CCFs are very similar from Monday to Friday, while a clear change in phase and amplitude appears on Saturday and Sunday. The stability of the short-period CCFs during weekdays proves that anthropogenic noise can indeed be recorded coherently over several kilometers. We can therefore conclude that the reconstruction of the short-period part of Green's function is limited only by the non-uniform distribution of short-period noise sources, and not by attenuation. Furthermore, this observation suggests that the anthropogenic sources are not only localized in space but also repetitive in time.

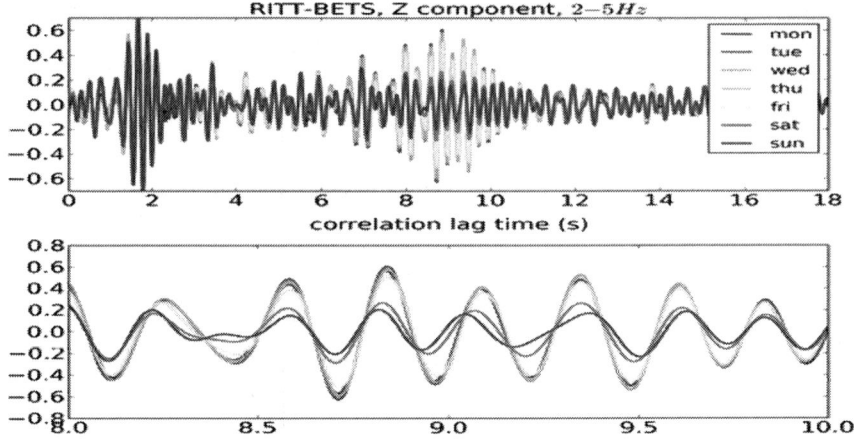

Figure 8: Vertical component cross-correlation functions measured between stations RITT and BETS and band-pass filtered between 0.2- and 0.5-s periods. The waveforms are averaged separately depending on the day of the week over 2 years. Bottom: zoom between 8 and 10 s of correlation lag time.

To refine our understanding of how the variability of anthropogenic sources impacts the correlations, we estimate the variability of the correlation waveform as a function of the local time of the day, restricting the analysis to weekdays (Monday to Friday). To do so, we first isolate the seismic noise recorded from Monday to Friday, then we compute the CCFs in 5-min windows. Finally, we stack the 5-min-long CCFs separately depending on the local time of the day over the whole acquisition period (for instance, we compute all the CCFs of the noise recorded at two stations every working day between local times 08:00 and 08:05 a.m. and stack them). The results for the RITT-BETS case are presented in Figure 9 (left side). Differences can be seen between day and night, with sudden changes of the waveform at 5 a.m. and 9 p.m. (horizontal dashed lines). A very energetic arrival is observed between 5 a.m. and about 7 a.m. (thick black arrow and closeup circle on Figure 9). Its origin has not yet been identified, but this arrival could be caused by a sudden increase of the traffic every morning at the same time along the same roads or by regular start-up times of machines in neighboring industries. The same energetic arrival between 5 a.m. and 7 a.m. is observed for other station pairs but at different correlation lag times, indicating that it is probably caused by a unique localized source. Some arrivals of the early part of the CCF (thick white arrows on

Figure 9) can only be seen when the anthropogenic activity decreases (during the nighttime and lunchtime) and could be due to natural (wind related) or continuous anthropogenic noise sources (e.g., pumps or industries in continuous operation). Given the chosen period range (0.2 to 0.5 s), the distance separating the stations (2.8 km), and an estimated group speed ranging from about 0.18 to 0.3 km/s, we expect fundamental-mode Rayleigh waves to occur between 9 and 16 s of correlation lag time. Earlier arrivals observed throughout the day may be caused by clustered noise sources that are not aligned with the axis formed by the pair of stations. Such sources generate wave fields that reach the two stations with delay times shorter than the Rayleigh wave propagation time from one first station to the other. Interpreting these phases as part of Green's function would lead to an over-estimation of the group speed between the stations.

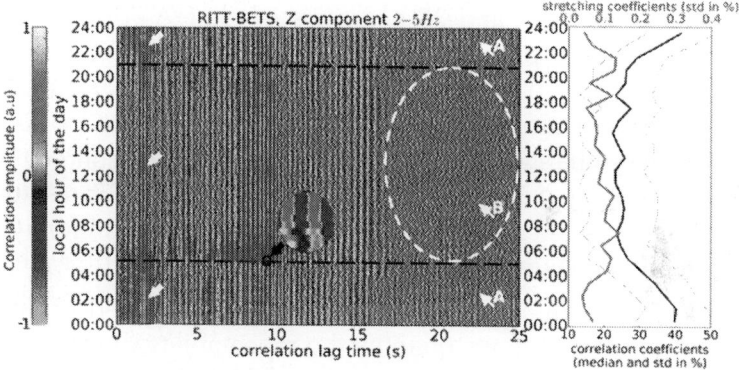

Figure 9: Evolution of the vertical component CCF waveform and the coda repeatability with local time. Left side: RITT-BETS CCFs, obtained from Monday to Friday, band-passed between 0.2 and 0.5 s and stacked separately for all the 5-min time slots of the day over the 2 years of available data. The stations are 2.8 km apart. The color bar corresponds to the amplitudes of the CCFs in arbitrary units. The thick black arrow indicates an enlarged part of the figure displayed in the closeup circle. Right side: temporal analysis of the coda (17 to 30 s of correlation lag time) performed separately for each hour of the day using a 30-day sliding window. The black curves show the median and standard deviation of the correlation coefficients measured between the coda and its reference. The red curve shows the standard deviation of the stretching coefficients measured before the drilling activities. We interpret these as the lowest detectable speed variations.

The late part of the correlation function (coda, e.g., after 16 s in the RITT-BETS case; Figure 9, left side) results from diffuse wave fields recorded coherently at both stations (seismic waves refracted on scatterers while traveling from one station to the other). Sens-Schönfelder and Wegler ([2006]) proposed to study the variability of the CCF coda over time to highlight velocity changes within the medium. This technique first establishes a reference coda by averaging the CCFs on a time span over which the medium is assumed to be invariant. Then, the CCFs computed on a sliding window are compared to this reference in order to identify infinitesimal variations (waveform stretching) of the coda. Obviously, the method requires the coda to be extremely repeatable so that any modification in its waveform can be attributed to changes into the medium.

We observe that the coda part of the CCF seems more stable during the night than during the day (Figure 9, left side). The early part of the coda (i.e., between 17 and 25 s, arrows labeled 'A') displays similar waveforms from 10 p.m. to 4 a.m., while no coherent phases can be seen in this part of the coda from about 8 a.m. to 9 p.m. (white dashed circle and arrow labeled 'B'). We infer that the positions of diurnal sources change more than those of nocturnal sources within our time resolution of 5 min. The daytime coda of the CCFs results from illuminating the scatterers around the station pair in a randomly time-varying manner, making it less repeatable.

In order to determine which part of the day is most suited for temporal analysis of the medium, we quantify the repeatability of the coda over time using the techniques of Sens-Schönfelder and Wegler ([2006]) and Brenguier et al. ([2008]). We conduct this analysis separately for each hour of the day. We first calculate 24 reference CCFs by averaging the CCFs separately as a function of local hour over the whole acquisition period. Then, for each local hour, we estimate how the coda computed over a 30-day sliding window resembles its reference CCF. Finally, for each local hour, we obtain a set of stretching coefficients (SCs) and their corresponding correlation coefficients (CCs). Medians and standard deviations of the CCs are used as indicators of coda repeatability (for instance, a value of 100% ± 0% would correspond to a coda that always matches its reference whatever the position of the 30-day window). Results obtained in the RITT-BETS case, using the 17- to 30-s coda filtered between 0.2 and 0.5 s (2 to 5 Hz), are presented in Figure 9 (right side, black curves). The coda is confirmed to be more stable at nighttime (up to 40% ± 10% correlation between the coda and its reference). The standard deviation of the SCs obtained before the first drilling (190 days) is also displayed as a function of local hour (Figure 9, right side, red curve). This curve is used as an indicator of the smallest detectable relative speed variation (v/v) that could be observed using our data set with a temporal resolution of 30 days. The detectable speed variation is about 0.1% during the day and 0.05% at night.

CONCLUSIONS

In this work, we benefited from the high station density available close to the two geothermal sites of Soultz-sous-Forêts and Rittershoffen and the long duration of available data (up to 4 years). We propose a detailed analysis of the seismic noise recorded in the area. The period range of interest is constrained by the dimension of the targeted structures. Based on the estimated seismic velocity model of the area, investigating the first 5 km of the crust requires working at periods between 0.2 and 5 s, which include seismic noise that has various origins and properties. At periods above 1 s, the secondary micro-seismic peak dominates the signal. This peak is characterized by a strong directivity (approximately 300° back azimuth) in good agreement with its origin (northern part of the Atlantic Ocean). At periods below 1 s, the noise has clear daily and weekly periodicities, which indicate its anthropogenic origin. The spatial analysis of this noise reveals that the sources are numerous but clustered around a few zones that roughly correspond to the densely populated villages of the area.

We compute CCFs for all station pairs of the network and analyze two of their properties. Firstly, we examine how the CCFs resemble Green's functions in terms of signal-to-noise ratio (SNR), dispersive behavior, and phase. This property of the CCFs is required for modeling the geographical distributions of seismic velocities (tomography), which will lead to better knowledge of the geological structures and characterization of the geothermal reservoir. At periods above 1 s, the SNR is low and the spatial distribution of the (mainly anthropogenic) noise sources limits our ability to reconstruct Green's function, making dispersion measurements difficult. At periods below 1 s, the SNR is higher. However, the high directivity of the noise at these periods affects the phase of the CCFs in a way that cannot be neglected. We expect CCFs to provide reliable information about distributions of seismic velocities inside the reservoir only if accurate knowledge of the noise directivity and rigorous estimates of errors induced on the phase are taken into account. This issue will be addressed in a forthcoming study.

Secondly, we analyze the stability of the correlation functions in time. This property is commonly used to follow the temporal variations of seismic velocities at depth and does not require a perfect match between the CCF waveform and the true Green's function. This technique is expected to provide information about changes that could occur inside the reservoir due to geothermal activities (relative displacement of scatterers induced by pressure variations, thermal fluctuations, variations of the fluid content, etc.). We show that high-frequency noise (0.2 to 0.5 s) due to anthropogenic activity is more stable/repeatable at night. The nocturnal noise sources, although non-uniformly distributed, seem to be more stable in space and time, making nocturnal CCFs more suited for temporal analysis. With a time resolution of 30 days, we estimate the smallest detectable relative phase speed variation to be about 0.05% to 0.1%. Future work will focus on the temporal variations of the medium induced by the operations conducted at the geothermal sites (drilling, injection/production tests, etc.).

AUTHORS' CONTRIBUTIONS

ML participated in data acquisition, performed the data processing and analysis and drafted the manuscript. JV designed and coordinated the study, and participated in data acquisition. JS and AM participated in the design and coordination of the study. JS also contributed to the data acquisition. All authors have read and approved the final manuscript.

ACKNOWLEDGEMENTS

This work has been published under the framework of the LABEX ANR-11-LABX-0050_G-EAU-THERMIE-PROFONDE and benefits from a funding from the state managed by the French National Research Agency as part of the Investments for the future program. ML is funded by Groupe Electricité de Strasbourg. We thank GEIE EMC, ECOGI and EOST for providing the data of the permanent network. The array equipment (ARIT and AKUL) was supplied by the SisMob component of the RESIF National Research Infrastructure. We thank the Geophysical Instrument Pool Potsdam (GFZ) for providing temporary stations, as well as E. Gaucher (KIT), V. Maurer (ES-G), H. Wodling, H. Jund, and M. Grunberg (EOST) who deployed the stations and collected the data. We are grateful to the three anonymous reviewers for their constructive criticisms that greatly helped improve the content of this manuscript.

REFERENCES

1. Bensen GD, Ritzwoller MH, Barmin MP, Levshin AL, Lin F, Moschetti MP, Shapiro NM, Yang Y:Processing seismic ambient noise data to obtain reliable broad-band surface wave dispersion measurements. *Geophys J Int* 2007, 169(3):1239-1260. doi:10.1111/j.1365-246X.2007.03374.x

2. Bonnefoy-Claudet S, Cotton F, Pierre-Yves B: The nature of noise wavefield and its applications for site effects studies: a literature review. *Earth Sci Rev* 2006, 79(3–4):205-227. doi:10.1016/j.earscirev.2006.07.004

3. Brenguier F, Shapiro NM, Campillo M, Ferrazzini V, Duputel Z, Coutant O, Nercessian A:Towards forecasting volcanic eruptions using seismic noise. *Nat Geosci* 2008, 1(2):126-130.

4. Brenguier F, Clarke D, Aoki Y, Shapiro NM, Campillo M, Ferrazzini V: Monitoring volcanoes using seismic noise correlations. *Comptes Rendus Geosci* 2011, 343(8–9):633-638. doi:10.1016/j.crte.2010.12.010

5. Bussat S, Kugler S: Offshore ambient-noise surface-wave tomography above 0.1 Hz and its applications. *Leading Edge* 2011, 30(5):514-524. doi:10.1190/1.3589107

6. Calò M, Dorbath CC: Different behaviours of the seismic velocity field at Soultz-sous-Forêts revealed by 4-D seismic tomography: case study of GPK3 and GPK2 injection tests. *Geophys J Int* 2013, 194(2):1119-1137.

7. Calò M, Dorbath C, Cornet FH, Cuenot N: Large-scale aseismic motion identified through 4-DP-wave tomography. *Geophys J Int* 2011, 186(3):1295-1314.

8.   Calò M, Kinnaert X, Dorbath C: Procedure to construct three-dimensional models of geothermal areas using seismic noise cross-correlations: application to the Soultz-sous-Forêts enhanced geothermal site. *Geophys J Int* 2013, 194(3):1893-1899. doi:10.1093/gji/ggt205

9.   Campillo M, Sato H, Shapiro NM, van der Hilst RD: Nouveaux Développements de L'imagerie et Du Suivi Temporel à Partir Du Bruit Sismique. *Comptes Rendus Geosci* 2011, 343(8–9):487-495. doi:10.1016/j.crte.2011.07.007

10. Derode A, Larose E, Campillo M, Fink M: How to estimate the Green's function of a heterogeneous medium between two passive sensors? Application to acoustic waves. *Appl Phys Lett* 2003, 83(15):3054-3056. doi:10.1063/1.1617373

11. Derode A, Larose E, Tanter M, de Rosny J, Tourin A, Campillo M, Fink M: Recovering the Green's function from field-field correlations in an open scattering medium (L). *J Acoust Soc Am* 2003, 113(6):2973-2976. doi:10.1121/1.1570436

12. Dziewonski A, Bloch S, Landisman M: A technique for the analysis of transient seismic signals. *Bull Seismol Soc Am* 1969, 59(1):427-444.

13. Froment B, Campillo M, Roux P, Gouédard P, Verdel A, Weaver RL: Estimation of the effect of nonisotropically distributed energy on the apparent arrival time in correlations. *Geophysics* 2010, 75(5):SA85-SA93. doi:10.1190/1.3483102

14. Gaucher E, Maurer V, Wodling H, Grunberg M: *Towards a dense passive seismic network over Rittershoffen geothermal field. 2nd European Geothermal Workshop.* KIT, EOST, Strasbourg, France; 2013.

15. Gouédard P, Roux P, Campillo M, Verdel A: Convergence of the two-point correlation function toward the Green's function in the context of a seismic-prospecting data set. *Geophysics* 2008, 73(6):V47-V53. doi:10.1190/1.2985822

16. Groos JC, Ritter JRR: Time domain classification and quantification of seismic noise in an urban environment. *Geophys J Int* 2009, 179(2):1213-1231.

17. Gutenberg B: On microseisms. *Bull Seismol Soc Am* 1936, 26(2):111-117.

18. Hadziioannou C, Larose E, Coutant O, Roux P, Campillo M: Stability of monitoring weak changes in multiply scattering media with ambient noise correlation: laboratory experiments. *J Acoust Soc Am* 2009, 125(6):3688-3695.

19. Kedar S, Longuet-Higgins M, Webb F, Graham N, Clayton R, Jones C: The origin of deep ocean microseisms in the North Atlantic Ocean. *Proc Royal Soc A Math Phys Eng Sci* 2008, 464(2091):777-793. doi:10.1098/rspa.2007.0277

20. Larose E: *Diffusion multiple des ondes sismiques et expériences analogiques en ultrasons*. Université Joseph-Fourier - Grenoble I, France; 2005.

21. Larose E, Roux P, Campillo M: Reconstruction of Rayleigh-Lamb Dispersion Spectrum Based on Noise Obtained from an Air-Jet Forcing. *The Journal of the Acoustical Society of America* 2007, 122(6):3437-44.

22. Lin F-C, Tsai VC: Seismic interferometry with antipodal station pairs. *Geophys Res Letts* 2013, 40(17):4069-4613. doi:10.1002/grl.50907

23. Lin F-C, Tsai VC, Schmandt B, Duputel Z, Zhan Z: Extracting seismic core phases with array interferometry. *Geophys Res Lett* 2013, 40(6):1049-1053. doi:10.1002/grl.50237

24. Lobkis OI, Weaver RL: On the emergence of the Green's function in the correlations of a diffuse field. *J Acoust Soc Am* 2001, 110(6):3011-3017. doi:10.1121/1.1417528

25. Longuet-Higgins MS: A theory of the origin of microseisms. *Phil Trans R Soc A* 1950, 243(857):1-35. doi:10.1098/rsta.1950.0012

26. McNamara DE, Buland RP: Ambient noise levels in the continental United States. *Bull Seismol Soc Am* 2004, 94(4):1517-1527.

27. Mordret A, Landés M, Shapiro NM, Singh SC, Roux P, Barkved OI: Near-surface study at the Valhall oil field from ambient noise surface wave tomography. *Geophys J Int* 2013, 193(3):1627-1643. doi:10.1093/gji/ggt061

28. Obermann A, Planès T, Larose E, Campillo M: Imaging preeruptive and coeruptive structural and mechanical changes of a volcano with ambient seismic noise. *J Geophys Res Solid Earth* 2013, 118(12):6285-6294. doi:10.1002/2013JB010399

29. Pedersen HA, Krüger F: Influence of the seismic noise characteristics on noise correlations in the Baltic shield. *Geophys J Int* 2007, 168(1):197-210.

30. Poli P, Campillo M, Pedersen H: Body-wave imaging of Earth's mantle discontinuities from ambient seismic noise. *Science* 2012, 338(6110):1063-1065. doi:10.1126/science.1228194

31. Rost S, Thomas eC: Array seismology: methods and applications. *Rev Geophys* 2002, 40(3):1008. doi:10.1029/2000RG000100

32. Roux P, Kuperman WA: and the NPAL Group: Extracting coherent wave fronts from acoustic ambient noise in the ocean. *J Acoust Soc Am* 2004, 116(4):1995-2003. doi:10.1121/1.1797754

33. Roux P, Sabra KG, Kuperman WA, Roux A: Ambient noise cross correlation in free space: theoretical approach. *J Acoust Soc Am* 2005, 117(1):79-84. doi:10.1121/1.1830673

34. Sabra KG, Gerstoft P, Roux P, Kuperman WA, Fehler MC: Extracting time-domain Green's function estimates from ambient seismic noise. *Geophys Res Lett* 2005, 32(3):L03310.

35. Sabra KG, Gerstoft P, Roux P, Kuperman WA, Fehler MC: Surface wave tomography from microseisms in southern California. *Geophys Res Lett* 2005, 32(14):L14311.

36. Sabra KG, Roux P, Kuperman WA: Arrival-time structure of the time-averaged ambient noise cross-correlation function in an oceanic waveguide. *J Acoust Soc Am* 2005, 117(1):164-174.

37. Sens-Schönfelder C, Wegler U: Passive image interferometry and seasonal variations of seismic velocities at Merapi Volcano, Indonesia. *Geophys Res Lett* 2006, 33(21):L21302. doi:10.1029/2006GL027797

38. Sergeant A, Stutzmann E, Maggi A, Schimmel M, Ardhuin F, Obrebski M: "Frequency-Dependent Noise Sources in the North Atlantic Ocean.". *Geochem Geophys Geosyst* 2013, 14(12):5341-5353. doi:10.1002/2013GC004905

39. Shapiro NM, Campillo M: Emergence of broadband Rayleigh waves from correlations of the ambient seismic noise. *Geophys Res Lett* 2004, 31(7):L07614.

40. Shapiro NM, Campillo M, Stehly L, Ritzwoller MH: High-resolution surface-wave tomography from ambient seismic

noise. *Science* 2005, 307(5715):1615-1618. doi:10.1126/science.1108339

41. Stehly L, Campillo M, Shapiro NM: A study of the seismic noise from its long-range correlation properties. *J Geophys Res* 2006, 111(B10):B10306.

42. Weaver R, Froment B, Campillo M: On the correlation of non-isotropically distributed ballistic scalar diffuse waves. *J Acoust Soc Am* 2009, 126(4):1817-1826. doi:10.1121/1.3203359

43. Weaver RL, Hadziioannou C, Larose E, Campillo M: On the precision of noise correlation interferometry. *Geophys J Int* 2011, 185(3):1384-1392. doi:10.1111/j.1365-246X.2011.05015.x

44. Withers MM, Aster RC, Young CJ, Chael EP: High-frequency analysis of seismic background noise as a function of wind speed and shallow depth. *Bull Seismol Soc Am* 1996, 86(5):1507-1515.

Citations

CHAPTER 1

Wei Yu and Kamy Sepehrnoori, "Optimization of Multiple Hydrauli-cally Fractured Horizontal Wells in Unconventional Gas Reservoirs," Journal of Petroleum Engineering, vol. 2013, Article ID 151898, 16 pages, 2013. doi:10.1155/2013/151898.

CHAPTER 2

Jinhyung Cho, Sung Soo Park, Moon Sik Jeong, and Kun Sang Lee, "Compositional Modeling for Optimum Design of Water-Alternating CO_2-LPG EOR under Complicated Wettability Conditions," Journal of Chemistry, Article ID 604103, in press.

CHAPTER 3

Gustavo Gabriel Becerra, Célio Maschio;and, Denis José Schiozer, Petroleum Reservoir Uncertainty Mitigation through the Integration with Production History Matching, doi.org/10.1590/S1678-58782011000200005.

CHAPTER 4

Said A. al Hagrey, "2D Model Study of CO_2 Plumes in Saline Reservoirs by Borehole Resistivity Tomography,"International Journal of Geophysics, vol. 2011, Article ID 805059, 12 pages, 2011, doi:10.1155/2011/805059.

CHAPTER 5

Maoyuan Feng and Pan Liu, "Spillways Scheduling for Flood Control of Three Gorges Reservoir Using Mixed Integer Linear Programming Model," Mathematical Problems in Engineering, vol. 2014, Article ID 921767, 9 pages, 2014. doi:10.1155/2014/921767.

CHAPTER 6

Shijun Huang, Baoquan Zeng, Fenglan Zhao, Linsong Cheng, and Baojian Du, "Water Breakthrough Shape Description of Horizontal Wells in Bottom-Water Reservoir," Mathematical Problems in Engineering, vol. 2014, Article ID 460896, 9 pages, 2014. doi:10.1155/2014/460896.

CHAPTER 7

CHAPTER 8

Index